A Compilation of Ligno-cellulose Feedstock And Related Research for Feed, Food and Energy

A Compilation of Ligno-cellulose Feedstock And Related Research for Feed, Food and Energy

D. A. Flores

Library of Congress Control Number:		2012923494
ISBN:	Hardcover	978-1-4797-6536-2
	Softcover	978-1-4797-6535-5
	Ebook	978-1-4797-6537-9

This book was printed in the United States of America.

To order additional copies of this book, contact:
Xlibris Corporation
1-888-795-4274
www.Xlibris.com
Orders@Xlibris.com
123130

Contents

Dedication

Dedicated in Memory to Pamela A. D. Rickard, Biochemist and Biotechnologist, Chair of the Department of Biotechnology, School of Biological Technologies, the University of New South Wales, Sydney, P. P. Gray and N. W. Dunn, who helped pioneer research on ligno-cellulose and sugarcane waste to produce ethanol during the early '70s petrol crisis, and to P. L. Rogers and his research on 'wet' and 'dry' fermentation and to Emeritus Prof. Ron A. Leng, A. O., Ph. D., D. Rur. Sci., FASAP for his dedication to research on fibroin and byproducts as a cheap source of animal feed for livestock, for food and for energy, the University of New England, Australia.

Author Biography

The author has had a general background in animal nutrition, specifically, with ruminant livestock and which started by coursework and research on the subject of this book at the University of New South Wales, Sydney and was recently awarded a postgraduate, Ph.D. degree, at the University of New England, Australia.

The author has in the past published on ensilage, protein digestion and intake, rec-DNA applications to low-quality feeds utilization and the improvement of temperate and tropical ensilage and rumen digestion with biotechnology.

The author is currently a web-based or Internet researcher and continues to research and publish, amongst others, in his area and specialty of low-quality feeds utilization and animal production. He resides in the municipality of Port Coquitlam, British Columbia, Canada.

Preface

This body of research is on the topic of biotechnology and ligno-cellulose as feedstock for animal livestock feeding, and which may also apply to feedstock for bioenergy, compiled from 2003 to 2008, and recently brought to date.

It is part of exploring new areas opening up in enzyme technology to breakdown ligno-cellulosic fibre, which have not yet here-to-fore been isolated, identified and characterized sufficiently for use 'as is', and with genetically modified organisms (GMOs) in ensilage for temperate and tropical climes and in the genetic engineering of DNA of GMO crops for animal feeding and their byproducts to improve utilization, for e. g., improving nutrient content of higher value (e. g. 'surrogate' proteins, water-soluble carbohydrate (WSC), lignin content) and controlling rate of degradability which leads to loss of nutrients during digestion (e. g. repression of protease activity, heat protection and use of tannins with proteins).

Enzyme technology has already shown promise with improving fermentation and nutritive value of silages. Eventually top dressing feeds with ensilage or 'as fed' with lignases that need further characterization (e. g. etherases and lyases) and including those from anaerobic species such as mesophilic, municipal activated waste sludge digestors and those resident in the rumen will also be brought into line along with other technologies although they are at an early stage.

The use of agro-industrial byproducts (AIBPs) (e. g. industrial food processing byproducts) have come to the fore again in developed countries like Japan, whilst feed residue byproducts including energy and protein concentrates and marginal land browse trees and shrubs and agricultural land systems for planting cash crops, animal fodders and food crops continue to be promoted because of their enormous potential for feed, livestock as capital, for food and for extra income in developing country settings.

Pre-treatment of ligno-cellulosic fibre either as feed or as has been researched as bioenergy feedstock is best done chemically (sulfur dioxide, SO2) and thermally (steam explosion, SE) to disrupt the ligno-cellulose complex and breakdown the ligno-hemicellulose bonds. This form of pre-treatment with co-generation of energy from biomass will become viable with time with mills brought up to function as bioenergy producers, including smaller co-operatives. The whole complex of producers or growers, their co-ops, transport lines or networks and milling stations or plants will have beneficial effects on employment and standards of living in rural areas in both the industrialized and developing world.

Recent approaches to feed pre-treatment, for e.g., with SO2/SE with ammoniation would act to disrupt the ligno-cellulose component of fibre and further breakdown lignin and supplement with N, pre-treatment with solid-substrate fermentation (SSF) with Basidiomycetes could be optimized further with O2-dependent lignase action and supplementation with microbial protein with minimized loss of organic matter (OM) and that pre-treatment with the "Yeast Bagasse Process", a process that has been already patented commercially, would also act to breakdown cellulose in the lignocellulose and supplement with single-cell protein (SCP).

The use of manipulation of rumen microbes shows promise with yeast and fungi as research continues including overcoming technical hurdles using fibrolytic rumen bacterial spp. with their identification, construction and testing as one of several approaches to modify rumen microbial digestion, is just beginning.

D. A. Flores, Ph.D.
Port Coquitlam, B.C. Canada

Chapter 1

ANAEROBIC LIGNINOLYSIS—
AS A RESEARCH GOAL.

The Problem and the Potential.

The process of lignification results in the deposition of phenolic derivatives in the plant cell wall material of forages which can be a barrier to accessibility by microbial enzymes in the digestion of forage feed material.

Alkali can be used, as in tropical settings, to delignify feed material although the biological route using microbial species in ensilage and in the rumen with digestion can be another potential approach that is being proposed here. The basis of the anaerobic process of lignin breakdown or ligninolysis has yet to be elucidated in order to apply it to processes in ensilage and the rumen.

There is great potential in applying the biological process of ligninolysis to feed material with ensilage and in ruminal digestion, which may increase the availability of energy from the forage of the animal's diet. This will of course depend on the degree to which lignin is depolymerized and exposure of the underlying polysaccharides under the biological conditions of ensilage or rumen digestion. There are also technical hurdles yet to be overcome with use of rumen modified organisms and if they subsist ecologically.

It is estimated alone without energy pressures on biomass supplies from crops that rice straw supplies have a comparative worth of several billions USD$ by economists as a commodity as a significantly upgraded feed by-product from being an already strategic staple food crop in Asia.

It is thus proposed that anaerobic ligninolysis be presented and assessed for its potential with more research proposed to bring about the major developments needed for its application to animal feeds.

Technical Research or Background.

Lignin's chemical structure of phenylpropanoid units may be linked in the polymeric form by various bonding configurations. These are illustrated in Fig. 1 following (taken from L. Wallace et al., 1983). Various aromatic compounds

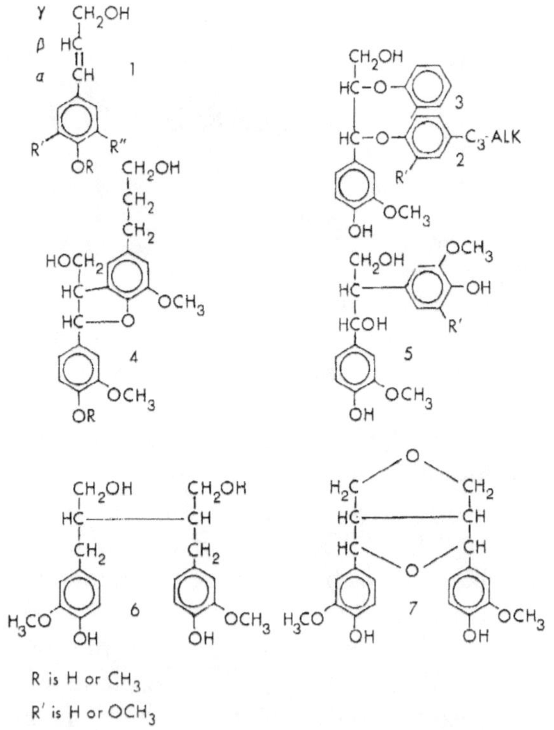

R is H or CH_3

R' is H or OCH_3

Fig. 1. Chemical structures of lignin.

(Taken from L. Wallace, A. Paterson, A. McCarthy, U. Raeder, L. Ramsey, M. MadDonald, R. Haylock and P. Broda (1983). The Problem of Lignin Biodegradation. Biochem. Soc. Symp. 48; 87-95.)

that are notable related to anaerobic metabolism of lignin include vanillic acid, dehydrodivanillate (DDV), ferulic acid, sinapic acid and coumaric acid. These compounds have been degraded or fermented by species found in the rumen or by protoplasts (D. E. Akin, 1980, W. Chen et al., 1988). *Pseudomonas spp.* have been shown to degrade vanillic acid as well as other lignin-related or model compounds (B. F. Taylor, 1983). There have been isolates from soil, composts and pulp mill effluent treatment plants that degrade lignin-related compounds. Some of the catabolic genes in plasmids have been transferred to *Pseudomonas putida* and *P. aeruginosa* (L. Wallace et al., 1983).

Colberg and Young (1982) prepared [14-C] lignin-labeled lignin by extracting lignin from cut twigs of Douglas fir grown with L-[U-14-C] phenylalanine using NaOH and heated into a slurry with supernatant separated and tested for any glucose with the extract consisting of molecular weight size fractions of soluble intermediates. The extract was then seeded with innocula from anaerobic mesophilic digestor fed waste-activated sludge. After 30 days, the original elution profile, from a Sephadex column, quantified by scintillation counting, which consisted of peaks in the 1400, 700 and 300 MW range, was reduced to eight peaks including the original 700 and 300 peaks with additional 900, 400, 200 and <200 MW peaks. The smaller 3 peaks represented the MW size range of single ring and smaller compounds. Unfortunately, samples were too dilute to determine spectrophotometrically ring fission. Further to this, collection of headspace gaseous carbon dioxide and methane and scintillation counting accounted for 13-18% of the original 14-C activity in a ratio of 2:3 demonstrating the mineralization of lignin.

The notable study above which involves oligomers of lignin, termed oligo- lignols, demonstrates lignin breakdown and intermediates from its breakdown separated out. Once further isolated, their molecular structures can be characterized as intermediates useful in describing the organic chemical basis to the biochemical breakdown process and to indicate the associated enzymatic activities that may be also isolated to characterize lignin breakdown in microbial metabolism. The enzymes, once isolated, can be used to probe or isolate genes responsible for ligninolysis in various microbial sources to be used for cloning in host species where ligninolysis is used for treating or breaking down lignin as has been suggested here with the ensilage process or digestion in the rumen. Further to this there might also be physico-chemical factors to the process that need further elucidation.

Anaerobic attack, with the degradation of lignified tissues, has been characterized by electron microscopy in the rumen with rumen isolate, strain 7-1, with plant tissues of the sclerenchyma in leaf blades and parenchyma in stems (D. E. Akin, 1980). This represents another potential source of genes that can be cloned involved with the anaerobic breakdown of lignin.

Another study differentiated the activities of bacterial and fungal degradation in rumen fluid measuring lignin oxidation products using radio-labelled lignin and characterizing plant tissues with electron microscopy (D. E. Akin and R. Benner, 1988).

In regards to the ensiling process, lignin-related enzymes and their genetic sources will have to be adapted first to the functional pH range from 7.0 to 3.8-4.0, as is achieved with most wet, well-fermented silages.

There is also the time of incubation factor where we assume that lignin degradation would be sufficient over a few weeks, the ensiling mass stabilizing over a few weeks, and in the rumen over a few days or during the retention time for feed in the rumen.

The Practical Benefits to Delignification.

The treatment of feed residues that are highly lignified with alkali will be used here as a practical example of the benefits that result from delignification or treatment of the lignin component of feed material. Alkali delignify some of the cell wall material by breaking hemicellulose-lignin bonds and may also break some bonds within the lignin molecule reducing its molecular weight; the effective delignification of the feed material then facilitates the penetration of microbial enzymes in the digestive process (P. T. Doyle et al., 1986).

With suitable moisture content and temperature conditions, microbes with ureases are capable of degrading urea with the formation of ammonium compounds such as ammonium carbonate, bicarbonate and hydroxide which are the agents that can permeate the feed material and help delignify the substrate (P. T. Doyle et al., 1986).

An example of a major study with the use of urea to upgrade or treat the feed material as opposed to just feeding or supplementation demonstrates dramatic differences in animal performance as shown with growth rates of 217 g / head /day compared with 718 g / head / day; the improvement was associated with intake differences of 1.8 kg / 100 kg liveweight compared with 2.4 kg / 100 kg liveweight (J. B. Schiere and J. Wieringa, 1988).

Short Summary.

Lignification presents a barrier to microbial enzymes in the digestion of forage feed material. Alkali can be used to delignify feed material but the biological route using microbial species in ensilage and in the rumen is proposed. The basis of anaerobic ligninolysis has yet to be elucidated. The economic worth of significantly upgraded rice straw alone from this proposed process is in the billions USD$. The potential of the process in increasing the energy availability of forages will depend on the degree of depolymerization of lignin under the biological conditions with ensilage and in the rumen. There are technical hurdles with use of modified rumen organisms in order for them to subsist. More research is proposed to bring about the major developments for its application to animal feeds. The chemical structure of lignin are phenylpropanoid units linked by various bonding configurations. Various aromatic compounds that have been degraded or fermented by species found in the rumen and by protoplasts include vanillic acid, DDV, ferulic acid, sinapic acid and coumaric acid. *Pseudomonas spp.* have been shown to degrade vanillic acid as well as other lignin-related compounds; there have been isolates from soil, composts and pulp mill effluent that degrade lignin-related compounds. Some of the catabolic genes in plasmids have been transferred to *Pseudomonas putida* and *P. aeruginosa*. Colberg and Young (1982) using a [14-C] lignin-labeled extract consisting of molecular size fractions of intermediates consisting of 3 peaks on a Sephadex column with scintillation counting when seeded with anaerobic mesophilic digestor fed waste-activated sludge reduced to eight peaks after 30 days. The smaller 3 peaks represented the MW size range of single ring and smaller compounds. Also collection of gaseous carbon dioxide and methane demonstrated mineralization of lignin. This study indicates that intermediates once separated out can be further characterized and the organic chemical basis of the biochemical process described with associated enzymatic activities to characterize lignin breakdown in microbial metabolism. The enzymes once isolated can be used to isolate genes which can be cloned in host species for ensilage or rumen digestion. Anaerobic attack on lignified tissues have been characterized by electron microscopy with a rumen isolate *strain 7-1* with plant tissues. Another study differentiated the activities of bacterial and fungal degradation in rumen fluid using radiolabelled lignin and characterizing plant tissues by electron microscopy. In regards to ensilage, enzymes and their genetic sources will have to be adapted to the funtional pH range of 7.0 to 3.8-4.0 for wet, well-fermented silages. The time of incubation would have to be over a few weeks with ensilage and in the rumen over a few days during the rumen retention time for feed. To give an example of the benefits of delignification, the example of the use of alkali will be used. Alkali breaks hemicellulose-lignin bonds and breaks bonds within the lignin molecule reducing its molecular weight facilitating penetration of microbial enzymes in digestion. With suitable moisture and temperature conditions ureases degrade urea and forms ammonium carbonate, bicarbonate and hydroxide to delignify the substrate. A study which upgraded or treated the feed material as opposed to just supplementation showed growth rates of 718 g / head / day compared with 217 g / head /day and intakes of 2.4 kg / 100 kg liveweight compared to 1.8 kg / 100 kg liveweight.

References.

1. D. E. Akin. 1980. Attack on lignified grass cell walls by a facultative anaerobic bacterium. Appl. Env. Microb. 40: 809-820.
2. D. E. Akin and R. Benner. 1988. Degradation of polysaccharides and lignin by ruminal bacteria and fungi. Appl. Env. Microb. 54: 1117-1125.
3. W. Chen, K. Nagashima, T. Kajino, K. Ohmiya and S. Shimizu. 1988. Intergeneric protoplast fusion between Ruminococcus albus and an anaerobic recombinant FE7. Appl. Env. Microb. 54: 1249-1253.
4. P. J. Colberg and L. Y. Young. 1982. Biodegradation of lignin-derived molecules under anaerobic conditions. Can. J. Microbiol. 28: 886-889.
5. P. T. Doyle, C. Devendra and G. R. Pearce. 1986. Improving the feeding value through pretreatments. In: Rice Straw as a Feed for Ruminants. Pp. 54-89. IDP Canberra Australia.
6. J. B. Schiere and J. Wieringa. 1988. Overcoming the nutritional limitations of rice straw for ruminants: response of growing Sahiwal and local cross heifers to urea-upgraded and urea-supplemented straw. Asian-Australian J. of Ani. Sci. 1: 209-212.
7. B. F. Taylor. 1983. Aerobic and anaerobic catabolism of vanillic acid and some other methoxy-aromatic compounds by Pseudomonas sp. strain PN-1. Appl. Env. Microb. 46: 1286-1292.
8. L. Wallace, A. Paterson, A. McCarthy, U. Raeder, L. Ramsey, M. MacDonald, R. Haylock and P. Broda. 1983. The problem of lignin biodegradation. Biochem. Soc. Symp. 48: 87-95.

Chapter 2

AEROBIC FUNGAL LIGNINOLYSIS

The Problem and the Potential.

The biological process of aerobic fungal ligninolysis is used here to mean the microbial process that breaks down lignin under aerobic conditions involving aerobic fungi that degrade lignin in feed residues of low-quality also termed, in part, as solid state fermentation (SSF) which includes the breakdown of cellulose/hemi-cellulose. There is also the possible application of enzymes directly on feed which is also included here.

As there is an inadequate supply of good quality feed in developing countries and the feeding of fibrous residue feeds is low in available energy due to its highly lignified content, the application of SSF or the possible use of enzymes to degrade lignin may prove to make available feeds of better quality.

There is of course the ongoing question of when applications of the SSF process will become feasible in a developing country setting for animal feed processing. The SSF process has now been suggested for other industrial biotechnological bioprocess applications in a developing country setting (R. L. Howard et al., 2003).

The Basis of the Aerobic Ligninolytic Process.

SSF with fungi involves the process of ligninolysis in order to digest the polysaccharides by exposing them and cleaving them with fungal cellulases and hemicellulases (K. E. Hammel, 1997).

The underlying basis or chemical reactions of lignin degradation with white rot fungi involve the generation of lignin free radicals which are unstable and cause spontaneous reactions in lignin (K. E. Hammel, 1997). White rot fungi attack the lignin polymers, open aromatic rings and cleave carbon-carbon and aryl ether bonds causing the formation of intermediate polymers and low molecular weight fragments (F. Zadrazil et al., 1995).

Although the collection of identified enzymes: lignin peroxidases (LiPs), manganese-dependent peroxidases (MnPs) and lacasses cannot degrade lignin in themselves (F. Zadrazil et al., 1995), there is strong evidence to implicate these enzymes in the ligninolytic process.

LiPs are a group of ligninolytic enzymes that contain ferric heme and are oxidized by hydrogen peroxide to a two-electron deficient intermediate which in turn returns to its resting state by performing two one-electron oxidations of the next donor substrate (K. E. Hammel, 1997). LiPs oxidize a variety of non-phenolic lignin structures and other aromatic ethers (K. E. Hammel, 1997). LiPs can generate aryl cation radicals and bring about C-alpha-C-beta cleavage, a major route of ligninolysis in many white rot fungi (K. E. Hammel, 1997). The fragmentation pattern of model compounds and their aromatic ring to give the cation free radical that occurs in a mass spectrophotometer is similar to that when LiP acts on lignin structures (K. E. Hammel, 1997).

MnPs are a group of ligninolytic enzymes that use diffusable manganese stabilized by chelates such as glycolate or oxalate to oxidize phenolic substrates; first hydrogen peroxide oxidizes MnP and then MnP is reduced via the oxidation of Mn(II) to Mn(III) near the active site with Mn(III) returning to its resting state Mn(II) by the oxidation of the phenolic substrate (K. E. Hammel, 1997).

It has been pointed out that there might be a new role for lipid peroxidation with MnP that attack non-phenolic lignin structures as white rot fungi that lack LiP also degrade them and that cannot be attacked by Mn(III); MnP has been shown to promote the peroxidation of unsaturated lipids and which generate potent lipoxyradical intermediates and which cleaves non-phenolic lignin model compounds (K. E. Hammel, 1997).

Lacasses are copper-dependent oxidases that catalyze the oxidation of phenolic substrates to phenoxy radicals; after four electrons have been removed from the substrate, lacasse reduces molecular oxygen to water returning it to the native state (K. E. Hammel, 1997). Lacasse may also attack non-phenolic compounds with 'mediators' such as 2,2'-azino-bis-(3-ethylthiazoline)-6-sulfonate (ABTS) or 3-hydroxyanthranilic acid (3-HAA) with which lacasse can oxidize a wide range of aromatic lignin compounds; ABTS, 3-HAA and 1-hydroxybenzotriazole have been used in pulp bio-bleaching applications; 3-HAA is produced as an intermediate in the biosynthesis of a major fungal orange pigment, cinnabarinic acid (D. O. Krause et al., 2003).

Hydroxy free radicals, among the most reactive, are also produced by white rot fungi; it oxidizes not only lignin but also cellulose non-selectively unless generated in a highly site-specific manner within the lignin portion of the cell wall (K. E. Hammel, 1996).

To generate hydrogen peroxide required by LiPs and MnPs, white rot fungi produce oxidases one of which is glyoxal oxidase that reduces oxygen to hydrogen peroxide with the oxidation of 1-3 carbon aldehydes, natural extracellular metabolites, examples of which are glyoxal, methylglyoxal and glycoaldehyde (K. E. Hammel, 1997). Aryl alcohol oxidases also produce hydrogen peroxide using chlorinated anisyl alcohols which are extracellular metabolites which are oxidized producing hydrogen peroxide; this is differentiated from alkoxybenzyl alcohols which are substrates for LiP while chlorinated benzyl alcohols are not, separating the ligninolytic and hydrogen peroxide-generating pathways (K. E. Hammel, 1997). Sugar oxidases may also be involved in hydrogen peroxide generation (K. E. Hammel, 1997).

The Use of the SSF Process on Low-Quality Feeds.

The SSF process with lignocellulosic substrate using fungal cultures are characterized by the complete or almost complete absence of free liquid with water absorbed or in complexed form with the solid matrix and the substrate suitable for fungi requiring low water activity (R. L. Howard et al., 2003).

The white rot basidiomycetes are the typical species used for fermentation of feed material (F. Zadrazil et al., 1995). However, the levels of lignases produced by these fungi are too low and the Kcat of the lignases are very slow for sufficient biological pretreatment of feeds (D. G. Armstrong and H. J. Gilbert, 1991). It will be necessary to boost lignase production in species through a combination of tech-niques of mutagenesis and genetic engineering of fungi that have low ligninolytic activities. Genetic engineering of lignases is possible to increase specific activity and for pH and thermostability with the possible replacement of amino acid residues that affect binding at the catalytic site to increase activity or residues that are essential for protein stability under various pH and temperature conditions (R. L. Howard et al., 2003). Lignases are under secondary metabolite control (D. G. Armstrong and H. J. Gilbert, 1991). Genetic engineering should modify this to constitutive control.

There is also the problem of simultaneous action of white rot fungi to degrade cellulose/hemicellulose and lignin components rather than lignin selectively which produces little or no change in digestibility and even decreases the nutritive value of the substrate or feed (F. Zadrazil et al., 1995). *Coprinus* fungi is an example in point with treatment resulting in sizeable losses of dry matter which will require a reduction in treatment time to reduce the loss (R. M. Acharya, 1988). This suggests other approaches to be also used. *Coprinus cinereus* increases digestibility of barley straw from 45 to 55% with ten days growth at a pH of 7 to 9; there is a dry matter loss of 12%; when fungal growth continues for more than ten days, cellulose begins to be rapidly utilized (P. T. Doyle et al., 1986). The use of cel negative mutants allows for improvement of strains to modify lignin but not degrade cellulose (F. Zadrazil et al., 1995). Among the most promising white rot fungi are *Cyathus stercoreus* and *Dichomitus squalens* but both are characterized by a slow growth rate and requires competing flora to be suppressed (F. Zadrazil et al., 1995).

The low moisture content to obtain maximum yield for the product with fungi usually excludes the problem of bacterial contamination and the need for pasteurization (F. Zadrazil et al., 1995). With putrefaction by overgrowing bacteria pasteurization of the feed is required; this may be done using solar heat, sodium hydroxide, sulfuric acid, ammonia or the use of urea which generates ammonia through the autochthonous microorganisms on the feed material; also the two-stage process of ensiling used with SSF can be used to produce the acidity that retards putrefactive bacteria (D. O. Krause et al., 2003).

Urea on forage rapidly converts to ammonia with autochthonous microorganisms and inhibits putrefying bacteria; *Coprinus* can grow at pHs as high as 10 (D. O. Krause et al., 2003). The use of urea pretreatment with SSF with *Coprinus* would allow the double action of delignification by ammonia generated from urea and the SSF process and the supplementation of feed with non-protein nitrogen (NPN) and in addition the protein from fungi.

There is a need to carefully control the growth conditions of fermentation including the duration of incubation, temperature, pH, aeration, moisture content and nutrient addition (P. T. Doyle et al., 1986). It has been suggested that the use of SSF be used only in industrialized countries where there might be a shortage of cereal grains and protein concentrate feeds or where prices are prohibitive (F. Zadrazil et al., 1995). It has been suggested for developing countries that large-scale commercial units be developed to treat feeds and then making treated feeds available to small farms (R. M. Acharya, 1988). Access to these large, expensive plants and their control equipment still presents a problem to the small farmer (F. Zadrazil et al., 1995).

Finally, the use of cell-free lignases or enzymes rather than with innoculation by cultures of fungi as another approach to delignify feed residues has been referred to although more research has yet to be done (F. Zadrazil et al., 1995). There may be a need to further elucidate the physico-chemical process of delignification. The enzymes are too large to penetrate the intact cell wall of plants and perhaps lignases use low molecular diffusable reactive compounds to effect initial changes to the lignin substrate (R. L. Howard et al., 2003). It has been proposed that veratryl cation radical is such a low molecular compound or 'mediator' for LiP (D. O. Krause et al., 2003). Previously mentioned were 'mediators' for lacasse, ABTS, 3-HAA and 1-hydroxybenzotriazole used in pulp bio-bleaching applications. A recent pilot pulp and paper process (Lignozym ® process) that effectively delignifies wood combines lacasse with 'mediator' compounds with groups NO, NOH or HRNOH in a lacasse-mediator system whose use should be extended to forage (D. O. Krause et al., 2003).

Short Summary.

Aerobic fungal ligninolysis is the subject of this paper which is the process of ligninolysis by aerobic fungi with low-quality feed residues also termed the SSF process with the breakdown of cellulose/hemicellulose. Also included are the direct application of enzymes. With the lack of available feeds of good quality in developing countries and with available feeds of low-quality the use of SSF or possibly enzymes may prove to make available feeds of better quality. There is the question of when the SSF process will become feasible in developing countries although it is now being proposed for other bioprocess applications. SSF with fungi involves ligninolysis and to digest polysaccharides with cellulases and hemicellulases. White rot fungi involve generation of lignin free radicals that cause spontaneous reactions in lignin attacking polymers, opening aromatic rings and cleaving carbon-carbon and aryl ether bonds. Although the collection of enzymes: LiPs, MnPs and lacasses cannot in themselves degrade lignin there is evidence to strongly implicate them.

LiPs are a group of enzymes containing ferric heme and are oxidized by hydrogen peroxide which in turn oxidizes the substrate. LiPs oxidize a variety of non-phenolic lignin structures and other aromatic ethers. They can bring about C-alpha-C-beta cleavage. MnPs are a group of enzymes that use diffusable manganese with hydrogen peroxide oxidizing MnP and MnP oxidizing Mn(II) to Mn(III) which then oxidizes the phenolic substrates. There might be a role for lipid peroxidation with MnP that attacks non-phenolic structures. Lacasses are copper-dependent oxidases that oxidize phenolic substrates which then reduce oxygen to water. Lacasses may also attack non-phenolic compounds with 'mediators' and oxidize a wide variety of aromatic lignin compounds. ABTS, 3-HAA and 1-hydroxybenzotriazole have been used in pulp bio-bleaching. 3-HAA is produced as a metabolite by fungi. To generate hydrogen peroxide required by LiPs and MnPs, white rot fungi produce oxidases including glyoxal oxidase and aryl alcohol oxidases oxidizing extracellular metabolites as substrates. Sugar oxidases may also be involved. The SSF process is characterized by low water activity in the substrate suitable for fungi. White rot basidiomycetes are the typical species used. However, the levels of lignases produced are too low and the Kcat of lignases are very slow for sufficient biological pretreatment. Mutagenesis and genetic engineering to boost lignase production and genetic engineering of lignases to increase specific activity and for pH and thermostability by substitution of amino acid residues may be used. Lignases are under secondary metabolite control and should be made constitutive. There is a problem of lack of selectivity in attacking the cellulose/hemicellulose and lignin components. *Coprinus* is an example with sizeable losses of dry matter and will require a reduction in treatment time to reduce loss with other approaches. Use of cel negative mutants is also another approach. Among the most promising white rot fungi are *Cyathus stercoreus* and *Dichomitus squalens* although they present limitations. Although not usually required, pasteurization, to control overgrowing bacteria and putrefaction, may be required with various approaches. The use of urea-ammonia generation and alkaliphilic *Coprinus* would provide a double-action of delignification by ammonia and the SSF process with supplementation with NPN and in addition protein from fungi. There is a need to carefully control the growth conditions of the fermentation process. For developing countries, large-scale commercial units may be developed to treat feed but access to these large, expensive plants and their control equipment still presents a problem to the small farmer. Use of cell-free enzymes will have to further elucidate the physico-chemical process of delignification since the enzymes are too large to penetrate the cell wall and this may involve low molecular diffusable compounds.

A recent pilot pulp and paper process (the Lignozym ® process) uses lacasse with 'mediator' compounds with groups NO, NOH or HRNOH and should be extended to forage.

References.

1. R. M. Acharya. 1988. Keynote Address. In: Non-conventional Feed Resources and Fibrous Agricultural Residues. C. Devendra (Ed.). IDRC-ICAR Ottawa Canada.

2. D. G. Armstrong and H. J. Gilbert. 1991. The application of biotechnologies for future livestock production. In: Physiological Aspects of Digestion and Metabolism in Ruminants. T. Tsuda, Y. Sasaki and R. Kawashima (Eds.). Pp. 737-761. Academic Press. San Diego, USA.

3. P. T. Doyle, C. Devendra and G. R. Pearce. 1986. Improving the feeding value through pretreatments. In: Rice Straw as a Feed for Ruminants. Pp. 54-89. IDP Canberra Australia.

4. K. E. Hammel. 1996. Extracellular free radical biochemistry of ligninolytic fungi. New J. of Chemistry 20: 195-198.

5. K. E. Hammel. 1997. Fungal degradation of lignin. In: Drive by Nature: Plant Litter Quality and Decomposition. G. Cadisch and K. E. Giller (Eds.). Pp. 33-45. CAB International. Oxfordshire, U. K.

6. R. L. Howard, E. Abotsi, E. L. Jansen van Rensburg, S. Howard. 2003. Ligno- cellulose biotechnology: issues of bioconversion and enzyme production. African J. of Biotechnology 2: 602-619.

7. D. O. Krause, S. E. Denman, R. I. Mackie, M. Morrison, A. L. Rae, G. T. Attwood and C. S. McSweeney. 2003. Opportunities to improve fibre degradation in the rumen: microbiology, ecology and genomics. FEMS Microbiology Reviews 27: 663-693.

8. F. Zadrazil, A. K. Puniya and K. Singh. 1995. Biological upgrading of feed and feed components. In: Biotechnology in Animal Feeds and Animal Feeding. R. J. Wallace and A. Chesson (Eds.). Pp. 55-70. VCH Weinheim FRG.

Chapter 3

TROPICAL SUGAR GRASSES

The Problem and the Potential

Recently, new high-sugar grasses were developed from ryegrass from varieties sourced from the U.K./Belgium and in Nothern Italy (Anon., 2003a) with high levels of sugar or fructans (S. Young, 2004). Research is now underway to develop even higher sugar grass varieties (S. Young, 2004). High-sugar varieties presently contain about 50% more sugar than normal grasses (Anon., 2003b). Tests on their utilization in animals shows they are higher performing grasses than their normal counterparts (S. Young, 2004). Research has been undertaken at the Institute of Grassland and Environmental Research or IGER, for short, in Aberyswyth in Wales in the U.K. where the high-sugar grasses were developed (S. Young, 2004, Anon. (no date)).

The understanding of the regulation and the cloning of fructan metabolism genes in grasses may allow for the eventual genetic engineering of tropical grass varieties to improve them. In a 1986 report there was 1.2 million hectares (ha) of commercial grazing lands in the Philippines with a carrying capacity of 425,000 animal units (AUs) (1 AU=5 small ruminants) (O. O. Parawan and H. B. Ovalo, 1986). In the same report, there were 1 million ha of traditional coconut farms that could be converted to coconut-pasture operations (O. O. Parawan and H. B. Ovalo, 1986). Examples in the Philippines of shade-tolerant tropical grasses are *Puspalum conjugatum, Setaria palmifolia* and *Centrosema accresens* (O. O. Parawan and H. B. Ovalo, 1986). There are various methods including burning, over-grazing and plowing that can be used to upgrade grasslands such as those in lands to oversow them (F. A. Moog, 2002). These new grass varieties might also be selected for higher growth rate and better ground cover.

Sugarcane, one of the major crops grown in the Philippines, is another grass that is considered as one of the most efficient in terms of biomass production making it ideal for applying various technologies, including biotechnologies, to upgrade its utilization for animal production (eg. beef and dairy) such as boosting its sugar levels. Specifically, the green sugarcane tops of the plant, used as a by-product feed, should be upgraded using high-sugar boosting.

The Basis of Fructan Metabolism

There is research not only in the breeding of high-sugar grass varieties at IGER but also the study of the genes, some of which have been identified and mapped, that control sugar metabolism (Anon., 2004). How grasses control sugar levels in their tissues (S. Young, 2004) will eventually be understood with this knowledge.

At the Institute of Botany at the University of Basel in Switzerland, research is underway to further elucidate fructan metabolism using barley as a model system (Anon., 2004). Fructans, of the -2,6 and -2,1 linkage are formed, as proposed by researchers there, by two enzymes: sucrose-fructan 6-fructosyl transferase (6-SFT) and sucrose-sucrose 1-fructosyl transferase (1-SST) (Anon., 2004). The two enzymes from barley have been purified (Anon., 2004). They have found that the 1-SST transcript and the enzymatic activity is subject to rapid turnover thus making it a likely candidate as a 'pacemaker enzyme' of fructan biosynthesis (Anon., 2004). The regulation of fructan metabolism is at the level of transcription, hence, promoters are important in deciphering the events of regulation (Anon., 2004). There are now studies on the 6-SFT promoter in barley and 1-SST promoter in tall fescue in order to identify the key promoter elements and interacting protein partners that determine activity of these promoters (Anon., 2004). There are also investigations on soluble acid invertases, or SAIs, in barley that affect fructan metabolism (Anon., 2004). Other putative invertase, fructosyl transferase-like (like 6-SFT) and fructan degradation enzyme or FEH cDNAs from barley are also being characterized by expression in hosts (Anon., 2004).

Ruminal Utilization of Sugar Grasses

Fructans and other like water-soluble carbohydrates (WSC) may limit microbial protein synthesis in the rumen. In vivo studies using high-sugar grass and control grasses silages show increased intake, lower rumen ammonia-N concentrations and higher microbial N flow to the small intestine with greater WSC content in the diets (Anon., (no date)). The greater flow of microbial N or protein is associated with the higher g microbial N per kg N input in the diet, which is found on these diets (Anon., (no date)). The protein incorporated into milk protein from dietary protein is 30% vs 25% on the high vs lower WSC grass diet (S. Young, 2004) which is a reflection of the higher efficiency with which microbial protein N is produced and flows from the rumen. The greater intake is likely due in part to greater digestibility with higher WSC levels. With greater energy supply from organic matter digestion, there is a better balance between the N degraded for microbial protein synthesis and the required energy and thus greater microbial N flow.

The Production Benefits of Feeding Sugar Grasses

Production benefits from feeding high-sugar grasses are the improvement of animal welfare, with animals handling less ammonia and their feeding on a natural grass diet which is what they do naturally (S. Young, 2004). Benefits to the environment are in terms of decreased excretion of N in urine and feces with cows fed high-sugar grasses (S. Young, 2004). Farmers have the added benefit of feeding less costly supplements on higher-sugar grass diets and consumers would be attracted to buying animal products from animals produced under more natural conditions (S. Young, 2004). Hypothetical tropical high-sugar grasses might be suited for more intensive production. An example would be goat dairying with 15 head or more in the Philippines (O. O. Parawan and H. B. Ovalo, 1986).

Short Summary

With the development of ryegrasses sourced from the U.K./ Belgium and Northern Italy at the Institute of Grassland and Environmental Research in Aberystwyth, Wales in the U. K. containing higher levels of sugars or fructans and the eventual isolation and mapping of genes that control sugar metabolism or levels of sugars in plant tissues is the possibility of developing, by genetic engineering, tropical grasses high in sugars or fructans. There are commercial grazing lands in the

Philippines and traditional coconut farms which could be converted to coconut-pasture operations. Examples in the Philippines of shade-tolerant tropical grasses are *Puspalum conjugatum, Setaria palmifolia* and *Centrosema accresens*. There are techniques to oversow grasslands and effectively upgrade them with these hypothetical grasses. Fructan metabolism has begun to be elucidated as shown by one research group at the University of Basel in Switzerland who have identified two enzymatic activities: 6-SFT and 1-SST using barley as a model system with 1-SST as the likely 'pace-maker enzyme' of the two. Regulation of fructan metabolism is at the level of promoter control with studies on barley and tall fescue in order to identify the key elements and protein partners. SAIs or acid invertases are also implicated in fructan metabolism. Fructans and like WSC are known to increase the efficiency of microbial protein synthesis and N flows to the intestines. This is associated with increased intake and decreased ammonia levels. With these high-sugar grass diets, the animal handles less ammonia and animals feed on the more natural grass diet for the better welfare of the animal. The environment benefits in terms of decreased excretion of N by the animal. The farmer benefits from feeding less supplements. And the consumer benefits from buying animal products produced under more natural conditions. Hypothetical tropical high-sugar grasses might be suited for more intensive production as with goat dairying with 15 head or more in the Philippines.

References.

1. Anon. (no date). Work-package Number 3 (WP3) Improving the efficiency of rumen function. http://www.sweetgrassineurope.org/Public-Pages/research.htm.
2. Anon. 2003a. Integrated research leads to award-winning grass and oat varieties. Business. October issue: 5-6.
3. Anon. 2003b. Success tastes sweet for high-sugar grasses. http://www.iger bbsrc.ac.uk/News/PR/2003/22April03successtastessweetforhigh-sugargrasses.
4. Anon. 2004. V. J. N. Nagaraj. http://www.unibas.ch/bothebel/people/vinay/ vinay.htm
5. F. A. Moog. 2002. Country Pasture/Forage Resource Profile. Livestock Development Council.
6. O. O. Parawan and H. B. Ovalo. 1986. Integration of small ruminants with coconuts in the Philippines. In: Small Ruminant Production Systems in South and Southeast Asia. Pp. 269-279. C. Devendra (Ed.). IDRC Ottawa Canada.
7. S. Young. 2004. Sweet grass, contented cows. Internet document.

Chapter 4

LOW-LIGNIN FORAGES.

The Problem and the Potential.

It is well established that lignin, the substance that gives cell walls in plants their strength is a barrier to cell wall polysaccharide degradation and digestion by ruminants. There have been approaches to either remove lignin in plant cell walls (eg. chemically), proposed increase breakdown by rumen microbes in the rumen or reduce lignin content by genetic manipulation.

A common form of delignification by chemical means using urea-ammonia usually improves the digestibility of forage by 2-6% with other reports by 10% (P. T. Doyle et al., 1986). This improves digestive utilization by making more available cell wall polysaccharides for energy and for microbial protein synthesis. In one study comparing data of cattle on chemically treated vs supplemented straw-based diets with urea-ammonia there was a 169% increase in liveweight gain (g/d) by native cattle (C. Devendra, 1997).

The use of cereals straws as low-quality feed residues is one an area where manipulation of lignin content of forages can improve utilization. Other low-quality feeds in sugarcane-growing regions are the sugarcane by-products of bagasse and green sugarcane tops.

There have been recent developments which we will review on the down-regulation of lignin biosynthesis using biomolecular tools called sense or anti-sense repression, the latter of which uses anti-sense-oriented DNA (the strand of DNA that is opposite to that which is transcribed to mRNA) that is added to the cell which causes binding to mRNA deactivating it from being expressed.

Brown midrib mutant varieties of millet with 23% less lignin content show an increase in digestibility of 4% (D. J. Cherney et al., 1990) and show increased digestibility similar to chemically delignified forages and should represent the same potential with decreased lignin content.

It has been found in poplar plants that up to a 40% reduction in lignin content is tolerable and does not result in adverse effects on normal plant growth and development (R. Zhong et al., 2000). Perhaps up to an 18% increase (referring to previous figures) in digestibility can be achieved using both chemical pretreatment and down-regulated low-lignin forages making forages of low-quality of improved quality as a result.

Lignin Biosynthesis As It Relates to Down-Regulation.

The nature of the biosynthesis of the monolignols of lignin is becoming to be better understood (see Figure 1 for a schemata). Phenylalanine or tyrosine are the precursors which are converted to the first of three acids in the pathway, p-coumaric, ferulic and sinapic acid which are 3-hydroxylated and 3-methylated and 5-hyroxylated and 5-methylated, successively; some of the enzymes include caffeic acid O-methyltransferase (COMT) and caffeoyl coenzyme A O-methyltransferase (CCoAOMT) and ferulate-5-hydroxylase (F5H); these acids or hydroxycinnamates are transformed to their CoA derivatives by the enzyme 4-coumarate:CoA ligase (4CL) and to their aldehydes derivatives by cinnamoyl CoA reductase (CCR) and finally to their alcohol derivatives by cinnamyl alcohol dehydrogenase (CAD); these alcohols are p-coumaryl alcohol, coniferyl alcohol and sinapyl alcohol (D. O. Krause et al., 2003). The alcohols are dehydrogenatively polymerized to lignin (R. Zhong et al., 2000).

The role of 5-hydroxylation by F5H on various intermediates in the pathway results either in guaiacyl lignin pre-5-hydroxylation or syringyl lignin, post-5-hydroxylation. Changes in composition as brought about by down-regulation of lignin biosynthesis are accompanied by shifts in syringyl (S) to guaiacyl (G) lignin or their S/G ratio.

The O-methyl transferases responsible for methylation of 3- and 5-hydroxy intermediates are COMT and CCoAOMT. COMT acts on caffeic acid, derived from p-coumaric acid, and its CoA derivative caffeoyl CoA; it also acts on 5-hydroxyferulic acid, the 5-hydroxy derivative of ferulic acid; it also acts on 5-hydroxyconiferaldehyde and 5-hydroxyconiferyl alcohol which are successively derived from the CoA derivative of ferulic acid, feroloyl CoA. CCoAOMT acts on caffeoyl CoA as with COMT and on 5-hydroxyferoloyl CoA, the CoA derivative of 5-hydroxyferulic acid (R. Zhong et al., 2000, D. O. Krause et al., 2003).

The study of Zhong et al. (2000) was the transgenic analysis which studied the role of CCoAOMT in lignification of poplar plants. They used CCoAOMT cDNA isolated from a cDNA library and it was assumed that anti-sense expression of one CCoAOMT cDNA would repress all CCoAOMT genes in poplar; this cDNA

Figure 1. A Schemata for the Biosynthesis of Lignin in Poplar.

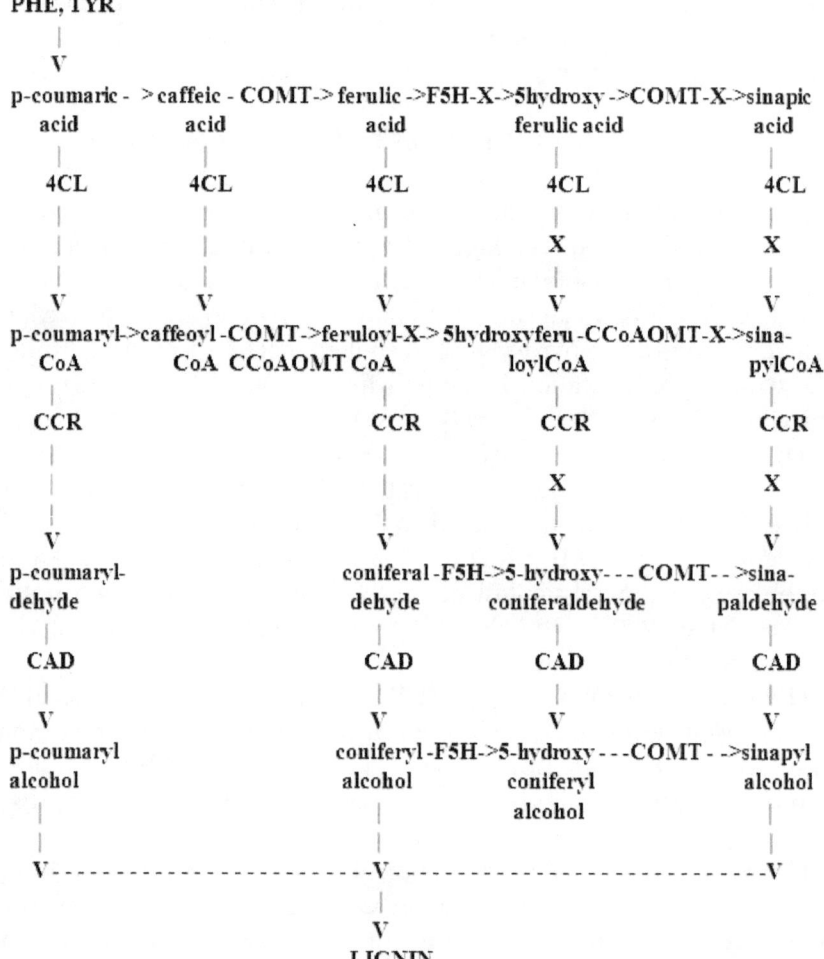

(Adapted from: D. O. Krause et al., 2003 and R. Zhong et al., 2000)
(Note: X-pathway blocked)

was inserted in the anti-sense orientation downstream from the cauliflower mosaic virus 35S promoter in the pBI121 binary vector to create pACoA; this was transformed into Agrobacterium which was used to infect popular stems; kanamycin resistance of transformants was used to select poplar transformants with anti-sense CCoAOMT cDNA; activity of CCoAOMT was then assayed from stem extracts of the plants; one transgene plant had a 30% reduction compared to the wild-type with the pBI121 vector; COMT was found to be normal as the wild-type; protein gel-blot analysis showed a dramatic decrease in CCoAOMT protein levels while COMT protein was the same.

A 70% reduction in CCoAOMT activity was associated with a 60% decrease in Klason lignin (R. Zhong et al., 2000). Pyrolysis mass spectrometry showed a decrease in guaiacyl and syringyl lignin units over polysaccharide (R. Zhong et al., 2000). Fourier transform infrared spectroscopy (diffuse reflectance) of wild-type transgenics showed a 56% decrease in lignin comparing absorbances for lignin and cellulose (R. Zhong et al., 2000). Using infrared bands indicating condensation and cross-linking of lignin showed a 82% reduction compared to the wild-type indicating less of it with reduced lignin content (R. Zhong et al., 2000).

Studies with poplar show that a 40% reduction in lignin only causes minor changes in vessel wall shapes and this mainly happened in the early stages of xylem development (R. Zhong et al., 2000).

An explanation put forth for the reduction in both guaiacyl and syringyl lignin is that the ferulate-5-hydroxylase (F5H) activity does not normally act on ferulic acid to form 5-hydroxyferulic acid and further 5-hydroxyferoloyl-CoA for 5-methylation by CCoAOMT but only on the 5-position of conferaldehyde and coniferyl alcohol which it exclusively does when presented with the three substrates; thus the action of COMT on 5-hydroxyferulic acid does not normally occur and neither does the action of CCoAOMT on 5-hydroxy-feroloyl CoA (see Fig. 1) (R. Zhong et al., 2000, D. O. Krause et al., 2003).

Studies show that COMT can efficiently 3-methylate caffeoyl CoA (from caffeic acid) to feroloyl CoA but it might be that CCoAOMT is present in much larger quantities than COMT in lignifying cells which cannot be compensated for by a reduction in CCoAOMT or COMT does not have in vivo access to caffeoyl CoA (R. Zhong et al., 2000). Feroloyl CoA which is 3-methoxylated is then 5-methoxylated with either coniferaldehyde or coniferyl alcohol to sinapyl alcohol (see Fig. 1) (R. Zhong et al., 2000, D. O. Krause et al., 2003).

A study in which anti-sense repression inhibited COMT activity while sense repression corepressed COMT activity decreased the S/G ratio by

40% and 90% and significantly improved cellulose degradability by 3.5% and 5.6%, respectively (M. A. Bernard-Vailhe et al., 1996).

The down-regulation of both CCR and CAD in the biosynthetic pathway results in a reduction in lignin levels with lignin similar to normal plant lignin; the plants also showed normal plant development (M. Chabannes et al., 2001). 4 coumarate:CoA ligase (4CL) responsible for the formation of CoA derivatives of the hydroxycinnamates when down-regulated in aspen showed similar features as previous (M. Chabannes et al., 2001).

A study with down-regulated CAD levels in alfalfa resulted in lower syringyl to guaiacyl ratios and lower syringyl + guaiacyl yield. Some of the down-regulated transgenic lines showed decreased in situ digestibility (M. Baucher et al., 1999).

Finally a study with mutant (ferulate-5-hydroxylase-deficient) *Arabidopsis* Where lignin concentration are the same as the wild-type and where syringyl-type lignin is absent did not show any differences in in vitro digestibility which contradicts findings where changes in composition of lignin is accompanied by changes in lignin digestibility (H.-J. G. Jung et al., 1999).

Brown Midrib Mutations.

Brown midrib (bmr) mutants, characterized by reduced lignin content and modified composition with improved digestibility (M. Baucher et al., 1999), have a red-brown coloration in their midrib and in their sclerenchyma and is associated with gene alteration in CAD as in the *bm1* mutant, in the caffeic acid/5-hydroxy ferulic acid O-methyltransferase (COMT) as in the *bm3* mutant in maize or an alteration in COMT and CAD activities as in the *bm6* mutant in sorghum (M. Baucher et al., 1999, D. O. Krause et al., 2003).

Lignin and Digestion of Plant Cell Walls.

Reduced content in lignin is probably associated with improved digestibility and results in improved access of rumen microorganisms and secreted enzymes to plant polysaccharides. However, it has been shown that the concentration of lignin in brown midrib mutants is poorly correlated with improved cell wall degradability (H.-J. G. Jung et al., 1999). There is a need to further explore the possibility of altering lignin content more subtly and the cell wall ultrastructure rather than overall lignin content.

The effects of composition on digestibility could in part be a reflection of inter-actions between lignin and cell wall polysaccharides and needs to be better understood. For example, it has been hypothesized that syringyl-rich lignin, being less branched, that is, having one less potential polymerization site, would protect a larger area of secondary cell wall from degradation than the more branched guaiacyl-rich lignin (H.-J. G. Jung et al., 1999)

Utilization of Low-Lignin Forage.

In feeding studies involving lambs feeding on normal and bmr pearl millet forage with 23% less lignin content, the digestibility was 4% higher on the bmr pearl millet; the first cutting showed no difference in intake although the second cutting had a higher intake for the bmr millet compared to the normal pearl millet, that is, 2.0 vs 1.5% of body weight (%BW); also, grazing lambs showed a marked preference for bmr pearl millet than the normal genotype spending more time on plots containing it (D. J. Cherney et al., 1990).

Short Summary.

Lignin is genetically modified for low lignin content in low-lignin forages. Down-regulation uses biomolecular tools of either sense or anti-sense repression using sense or anti-sense DNA in plant cells. Brown midrib mutants which have genetic defects in lignin biosynthesis are also a form of low-lignin forages. It is estimated as much as an 18% increase in digestibility can perhaps be realized with both the chemical pre-treatment and down-regulation of low-lignin forages. Lignin biosynthesis is becoming to be more fully understood with the three monolignols synthesized from phenylalanine to tyrosine as the first precursors. The monolignols are p-coumaryl alcohol, coniferyl alcohol and sinapyl alcohol. The three major acid intermediates are p-coumaric, ferulic and sinapic acid which correspond to the three alcohols respectively and are successively 3-methoxylated (ferulic acid) and 5-methoxylated (sinapic acid) to form guaiacyl lignin, pre-5-hydroxylation, or syringyl lignin, post-5-hydroxylation. The O-methyl transferases responsible for the methylation of 3- and 5-hydroxy intermediates are COMT and CCoAOMT. The down-regulation of CCoAOMT in poplar plants was recently studied which caused a 70% reduction in CCoAOMT and 60% decrease in lignin content. The transgenic analysis of the biosynthetic pathway reasoned that CCoAOMT which 3-methylates a CoA intermediate of caffeic acid, caffeoyl CoA, is present in much larger quantities normally than COMT and when repressed cannot be compensated for by COMT activity; thus both guaiacyl and syringyl lignin is reduced. Down-regulation of COMT, both CCR and CAD, 4CL, CAD and F5H involved with lignin biosynthesis have also been studied and so with brown midrib (bmr) mutants which result in either reduced lignin content or changes in composition and changes in digestibility although results were not consistent. It has been shown that the concentration of lignin is related to digestibility although it has to be better understood what it is in lignin content and the cells ultrastructure that determines this. Composition of syringyl and guaiacyl lignin may also determine the interactions between lignin and cell wall polysaccharides. Brown pearl millet showed a 23% less lignin content and increased digestibility of 4%; there was however a difference in intake only in the second cutting; lambs fed bmr pearl millet showed a preference for it when choosing between types.

References.

1. M. Baucher, M. A. Bernard-Vailhe, B. Chabbert, J.-M. Besle, C. Opsomer, M. V. Montagu and J. Botterman. 1999. Down-regulation of cinnamyl alcohol dehydrogenase in transgenic alfalfa (*Medicago sativa L.*) and the effect on lignin composition and digestibility. Plant Molecular Biology 39: 437-447.

2. M. A. Bernard-Vailhe, C. Migne, A. Cornu, M. P. Maillot, E. Grenet, J. M. Besle, R. Atanassova, F. Martz and M. Legrand. 1996. Effect of modification of the O-methyltransferase activity on cell wall composition, ultranstructure and degradability of trangenic tobacco. Journal of the Science of Food and Agriculture 72: 385-391.

3. M. Chabannes, A. Barakate, C. Lapierre, J. M. Marita, J. Ralph, M. Pean, S. Danoun, C. Halpin, J. Grima-Pettenati and A.-M. Boudet. 2002. Strong decrease in lignin content without significant alteration of plant development is induced by simultaneous down-regulation of cinnamoyl CoA reductase (CCR) and cinnamyl alcohol dehydrogenase (CAD) in tobacco plants. Tektran USDA Agricultural Research Service. http://www.nal.usda.gov/ttic/ tektran/ data/000012/64/0000126438.html.

4. D. J. Cherney, J. A. Patterson and K. D. Johnson. 1990. Digestibility and feeding value of pearl millet as influenced by the brown-midrib, low-lignin trait. Journal of Animal Science 68: 4345-4351.

5. C. Devendra. 1997. Crop residues for feeding animals in Asia: technology development and adaptation in crop/livestock systems. In: Crop residues in sustainable mixed crop/livestock farming systems. C. Renard (Ed.). Pp. 241-267. CAB International, Wallingford, Oxon

U.K. 6. P. T. Doyle, C. Devendra and G. R. Pearce. 1986. Improving the feeding value through pretreatment. In: Rice straw as a feed for ruminants. Pp. 54-89. IDP Canberra ACT Australia.

7. D. O. Krause, S. E. Denman, R. I. Mackie, M. Morrison, A. L. Rae, G. T. Attwood and C. S. McSweeney. 2003. Opportunities to improve fibre degradation in the rumen: microbiology, ecology and genomics. FEMS Microbiology Reviews 27: 663-693.

8. H.-J. G. Jung, W. Ni. C. C. S. Chapple and K. Meyer. 1999. Impact of lignin composition on cell-wall degradability in an *Arabidopsis* mutant. Journal of Food Science and Agriculture 79: 922-928.

9. R. Zhong. W. H. Morrison III, D. S. Himmelsbach, F. L. Poole II and Z.-H. Ye. 2000. Essential role of caffeoyl coenzyme A O-methyltransferase in lignin biosynthesis in woody poplar plants. Plant Physiology 124: 563-578.

Chapter 5

DOWN-REGULATED PROTEASES IN FORAGES.

The Problem and the Potential.

The down-regulation of plant proteases in forages as with grasses and legumes and with further protection of plant proteins aims to conserve protein and the supply of peptides and amino acids that stimulate ruminal microbial digestion and to increase protein bypass in the rumen from feeds.

Ecology of Down-regulation and Function in Plants.

There is the question as to whether inhibiting the action of plant proteases might have an ecological impact in terms of the natural degradation of plant matter in nature from forages although on the otherhand this may not be as serious an issue considering other contributing processes present in nature that brings about the degradation of plant material.

There are three major examples discussed in Zhu et al. (1999) by which plant proteases act in plant material. The first involves de novo synthesis of proteases in a process of controlled cell death whereby proteins are catabolized to amines and amino acids for use in other parts of the plant. The second involves the processing of proteins synthesized in the nucleus that are imported into organelles such as in chloroplast.

The third example acts to cauterize damage or injury to cells from infection by plant pathogens. Plant proteolysis must be highly regulated biochemically or by compartmentalization to protect the viability of co-existing plant proteins.

Plant Proteases and their Control.

It will be necessary to identify and isolate those proteases that are released during damage of plant material during grazing or harvesting of forage material and that act during rumen digestion or during ensilage, prior to feeding. It remains to be seen to what extent plant proteolysis play a role with grazed forages; although, it has been found that plant proteases contribute to initial stages of proteolysis of grazed herbages (W.-Y. Zhu et al., 1999). Also, in the rumen, given its well-adapted and high microbial proteolytic activity, that prior to ensiling, there is the opportunity of protecting proteins by wilting forages which may cauterize the process of plant proteolysis known to breakdown proteins to peptides and amino acids in the silo. It should be determined to what extent down-regulation of proteolysis would make a difference with feed material wilted in the field in rumen digestion.

Sense and anti-sense repression are recent molecular tools used to inhibit the enzymatic activities of certain enzymes in metabolic pathways where sense or anti-sense DNA is introduced into the cell, in the former case causing over-production of the copies of the enzyme and causing repression of its expression from end-products, while in the latter mRNA from the anti-sense copy of the gene would bind with normal mRNA deactivating transcription. It is hypothesized that anti-sense repression or down-regulation as it is also called will be used as it is not known as to whether the former form of repression with proteolysis with its general nature of breakdown of plant protein would be end-product repressed.

Short Summary.

Down-regulation of plant proteases may allow for increased supply of protein for ruminal digestion from forage increasing the supply of peptides and amino acids from its digestion and breakdown in the rumen for microbial fermentation. What plant proteases that generally act in the process of breakdown of plant protein in grazed or harvested herbage needs to be identified and isolated. The role of plant protease activity needs to be further investigated; it has been found that proteases play a role initially in grazed forage; also, it remains to be seen however to what extent plant proteins are degraded in the rumen despite their being wilted given the high, well-adapted microbial proteolytic activity in the rumen. Down-regulation, or anti-sense repression, it is hypothesized, might be a method of choice as opposed to sense repression.

References.

1. W.-Y. Zhu, A. H. Kingston-Smith, D. Troncoso, R. J. Merry, D. R. Davies, G. Pichard, H. Thomas and M. K. Theodorou. 1999. Evidence of a role for plant proteases in the degradation of herbage proteins in the rumen of grazing cattle. J. Dairy Sci 82: 2651-2658.

Chapter 6

UTILIZING EXOGENOUS FIBROLYTIC ENZYMES (EFES) WITH FEEDS

The Problem and the Potential

The effectiveness of approaches to using exogenous (cell-free) fibrolytic enzymes (EFEs) with plant cell walls in feeds, eg. EFEs together with alkali pretreatment, and the reduction in production costs of EFEs (Y. Wang and T. A. McAllister, 2002) leads to the suggestion of possible application of EFEs to feeds such as cereal straws such as in developing countries and may make the technology available in these countries in the future.

Among the enzymes in EFEs, the enzymatic mixture may include cellulolytic, hemicellulolytic, amylolytic and proteolytic enzymes (D. O. Krause et al., 2003).

Because a major issue facing EFE use is the variability in response found with EFEs (K. A. Beauchemin et al., 2002) we will now discuss factors that affect the effectivity of the process including the efficacy of EFEs from different sources, efficacy of EFEs applied by different methods, the efficacy of EFEs on different diets, the efficacy of EFEs and the level of animal production and improving not only the rate of digestion with EFEs but also improving their extent of digestion (Y. Wang and T. A. MacAllister, 2000).

Efficacy of EFEs from Different Sources.

EFEs that are marketed which are in the hundreds are primarily from four bacterial species: *Bacillus subtilis, Lactobacillus acidophilus, Lactobacillus plantarum* and *Streptococcus faecium*, three fungal species: *Aspergillus oryzae, Trichoderma reesei* and *Saccharomyces cerevisiae* and to a lesser extent fungal species *Humicola insolvens* and *Thermomyces spp.*; enzymatic preparations are referred to as cellulases or xylanases although invariably contain secondary activities of amylases, proteases and pectinases; enzymes are usually prepared to obtain specified levels of activity of cellulase and/or xylanase; multiple enzymes are required for either corresponding substrate and this may vary according to the microbial source and culture conditions (Y. Wang and T. A. McAllister, 2002). A middle approach of designing these enzymes of "one-size-fits-all" may have to be replaced with a more targeted or designer approach to ensure efficiency of feed enzymes. (K. A. Beauchemin et al., 2002).

Efficacy of EFEs Applied by Different Methods.

It has been found that direct application of EFEs to feed by spraying EFEs onto dry feed components (either forage or concentrate) prior to feeding is effective (Y. Wang and T. A. McAllister, 2002). One may speculate that this is due to increased enzyme stability, greater retention time with feed particulates in the rumen and preruminal hydrolysis of feed substrate by EFEs. The attachment of EFEs to feed particles would enhance the fact that the majority of the ruminal ecosystem's enzymatic activity is attached to the particulate phase rather than being in the fluid phase (Y. Wang and T. A. McAllister, 2002).

Extensive preruminal hydrolysis reduced ruminal microbial colonization despite increasing dry matter (DM) loss; this was not due to greater solubilization of sugars as released by enzymatic action; it is suggested that EFEs and ruminal microbes may compete for reaction sites on the substrate or some products of enzymatic action may be inhibitory to colonization (Y. Wang and T. A. McAllister, 2002).

It has been found that EFEs have a limited capacity to cleave phenolic com-pound mediated-lignin carbohydrate complexes (PC-LCCs) which may be inhibitory to microbial colonization involving esterified and etherified linkages causing the PC-LCC matrix to accumulate on the surface with EFE hydrolysis; limited hydrolysis prior to ruminal incubation may promote microbial growth by increasing availability of reducing sugars without the accumulation of substantive PC-LCCs at the feed surface (Y. Wang and T. A. McAllister, 2002). However, it is still not known to what extent preruminal hydrolysis by EFEs would be optimal for ruminal microbial digestion (Y. Wang and T. A. McAllister, 2002).

Studies investigating the effect of the moisture content of feeds (eg. silages, mixed rations or concentrate) speculate that high moisture feeds would have lower binding capacity so that EFEs would be dissolved into ruminal fluid; no studies to date have been done on feed moisture content per se (Y. Wang and T. A. McAllister, 2002).

It has been hypothesized that the greater the proportion the diet (eg. premixes, supplement or concentrate) is treated with enzymes, the greater the chances that enzymes endure in the rumen (K. A. Beauchemin et al., 2002).

Efficacy of EFEs on Different Diets.

Sources of EFEs are from aerobic fungi with pH optima 4.0-6.0; dairy and beef cattle have pHs below 6.0 during a significant portion of the day; there is an expected higher improvement in fibre digestion on high cereal grain diets; factors involved are substrate enzymatic specificity at the pH optima; also there is the lower occurrence of cellulolytic bacteria at lower pH (Y. Wang and T. A. McAllister, 2002).

The application of EFEs to forage diets (alfalfa and timothy) has also resulted in increased animal performance or growth rate (Y. Wang and T. A. McAllister, 2002).

Efficacy of EFEs and Level of Animal Production.

EFEs are likely to benefit animals fed for maximal productivity compared to those just at maintenance; with high producing diets, fibre digestion is often compromised or reduced due to lower ruminal pH and rapid transit times through the rumen (K. A. Beauchemin et al., 2002); this may be in the case of high cereal or grain diets and where cellulolysis is discouraged by lower pH in the rumen and where there is more rapid transit compared with less digestible, more fibrous diets; also with lower intake as % body weight (%BW), as with sheep, digestibility is higher compared with dairy cows and thus a higher improvement would be cap-tured with compromised or lower digestibility, in the latter case with dairy cows (K. A. Beauchemin et al., 2002).

The Extent vs. the Rate of Digestion.

It has been pointed out that EFEs would only increase the rate of digestion in the rumen and not the extent of digestion (D. O. Krause et al., 2003). In fact new or novel enzymatic activities would have to be introduced into the rumen to effect this increase (D. O. Krause et al., 2003). We will discuss here means which can be used together with EFEs to bring about in increase in extent of digestion. These have been suggested as those treatments or activities that increase ligninolysis to complement the activity of EFEs on cell wall digestion (D. O. Krause et al., 2003). They are urea or ammoniation, use of ligninolytic enzymes such as lacasses with mediators with NO, NOH and HRNOH groups and esterases and etherases that break polysaccharide-lignin bonds.

There has been an interest in the use of ammoniation with anhydrous or gaseous ammonia and aqeous ammonia or ammonium hydroxide in developed countries using cereal straws (P. T. Doyle et al., 1986). Ammonia levels significantly improve digestibility and may be due to effective solubilization of hemicellulose and the breakdown of lignin bonds especially at pHs above 10 (P. T. Doyle et al., 1986). However, urea is more readily available in less developed countries; pretreatments have resulted in 2-6% increases organic matter (OM) digestibility, some 10% (P. T. Doyle et al., 1986). This might not be seen to be as effective as ammonium hydroxide where the ammonium compounds formed at adequate moisture and temperature for microbes producing urease may be ammonium carbonate, bicarbonate and hydroxide and may only reach pHs of 8-9 which may not break the lignin bonds as effectively as a pH of 10 (P. T. Doyle et al., 1986). It is perhaps pertinent to discuss here possible means of improving urea-ammonia pretreatment of feeds. One would be to optimize level of application, moisture content and time for treatment with air tight conditions and use of urease additives (eg. from soybean) which reduces treatment times; a better understanding also is needed of which ammonium compounds are formed, the pH increase achieved and the reactions in urea-treated straw (P. T. Doyle et al., 1986).

Lacasses are one of the major lignases that have recently been investigated to have low molecular weight 'mediators' such as 2,2'-azino-bis-(3-ethylthiazoline)-6-sulfonate (ABTS) and 3-hydroxyanthrinilic acid (3-HAA) which allow it to degrade a wider range of aromatic lignin compounds; in fact, lacasse with such redox mediators as ABTS, 3-HAA and 1-hydroxybenzotriazole have been used in process-scale bio-bleaching of pulp; the use of low molecular mass mediators would be necessary as lignases are too large a molecule to penetrate the unaltered wood cell wall (D O. Krause et al., 2003). The Lignozym ® process which use physiological mediators with NO, NOH and HRNOH groups have been used in pilot pulp and paper making and should be applied to forages (D. O. Krause et al., 2003).

Finally when one examines the polysaccharide-lignin structure of plant cell walls there are various phenolic compounds that mediate lignin-carbohydrate complexes such as ferulic acid and p-coumaric acid cross bridges or linkages (Wang and Mc-Allister, 2002). Ester and ether bonds are formed (Krause et al., 2003) which can be broken by esterases (eg. acetylesterase and ferulic acid esterase) and etherases (Wang and McAllister, 2002). Adding additional activities to a cellulase/xylanase EFE preparation would introduce such novel activities and improve the extent of rumen digestion of cell wall polysaccharides.

Short Summary.

The effectiveness of approaches of using EFEs with plant cell walls and their reduction in cost producing diets has led to the suggestion of their possible application to feeds such as cereal straws such as in developing countries where this technology might be made available. Variability in results on the use of EFEs has been largely due to various identified factors. One is the efficacy of EFEs depending on the variability in the source of activity such as with cellulases and/or xylanases involving multiple enzymes which varies with microbial source and cultural conditions. A more targeted or designer approach may have to be required in the future to deliver on the effectiveness of EFEs. Another factor affecting the efficiency of EFEs is the method of application. It has been found that spraying EFEs onto dry components is more effective and this may be due to increased enzyme stability, greater retention time with feed particles and optimal preruminal hydrolysis of sugars on the feed surface. Also it has been hypothesized that the larger the proportion the diet is applied with EFEs, the greater time EFEs would endure in the rumen. Another factor influencing efficiency of EFEs is diet type. High cereal grain diets have the right pH range and the relative lack of cellulolytic bacteria at that pH range result in an improvement in fibre digestion. Forage diets also show an improvement in utilization with EFEs. Another factor here that affects the efficacy of EFEs is the level of animal productivity. On high con- centrate diets fibre digestion is more compromized due to lower ruminal pH and greater rapid transit times. Also at greater intake as %BW or less digestibility more of this loss is captured by use of EFEs. Finally, EFEs may increase only the rate of digestion in the rumen but may be limited by the extent as there would be a need for novel treatments or activities to be introduced to bring about breakage of polysaccharide-lignin bonds. Some of these are the application of alkali such as urea or ammoniation, lignases such as lacasses with mediators with groups NO, NOH, HRNOH and esterase and etherases which break the polysaccharide-lignin bonds.

References.

1. K. A. Beauchemin, D. Colombatto, D. P. Morgavi and W. Z. Yang. 2002. Use of Exogenous Fibrolytic Enzymes to Improve Feed Utilization by Ruminants. Journal of Animal Science 81 (E suppl.): E37-E47.
2. P. T. Doyle, C. Devendra and G. R. Pearce. 1986. Improving the Feeding Value through Pretreatments. In: Rice Straw as a Feed for Ruminants. Pp. 54-89. IDP, Canberra Australia.
3. D. O. Krause, S. E. Denman, R. I. Mackie, M. Morrison, A. L. Rae, G. T. Attwood and C. S. McSweeney. 2003. Opportunities to improve fibre degradation in the rumen: microbiology, ecology and genomics. FEMS Microbiology Reviews 27: 663-693.
4. Y. Wang and T. A. McAllister. 2002. Rumen microbes, enzymes and feed digestion (Internet document).

Chapter 7

SUGARCANE AS FEED.

The Problem and the Potential.

Sugarcane grown as a crop for sugar has in the past suffered in economic crisis with the low price for sugar; in the early 1980's, the price for sugar was US 27 cents a pound, in 1980, and plummeted to US 3.5 cents a pound in 1983 (E. T. Baconawa, 1986).

Sugarcane is a plant that has an advantage over other tropical grasses as a biomass producer and is perennial with both nutritional quantity and quality increasing and with optimum values reached at harvest intervals of between 12 to 18 months (T. R. Preston, 1986).

In the Philippines integrating livestock with the use of whole sugarcane as feed has been practiced beginning in the early 1980's (E. T. Baconawa, 1986) and would further diversify and add to the profit of growing sugarcane. In the Philippines, both beef cattle and dairy have been fed chopped, whole sugarcane in one farm for half the year with supplementation; dairy cows produce about 8.5 kg of milk/day during a lactation period of 260 days (E. T. Baconawa, 1986); and up to 0.84 kg/day average daily gains (ADGs) have been obtained with a diet of sugarcane tops (the aerial part of the sugarcane plant made of the green leaves, the bundle leaf sheath and the immature cane) supplemented with 1 kg rice polishings with cattle (M. R. Naseevan, 1986). Sugarcane bagasse can be used as a feed from sugar milling with treatment or fermentation (E. T. Baconawa, 1986). Molasses can be used in feeds for various livestock at 4-5% as the recommended rate. It is recommended at 8-10% use for rice straw which also makes the straw more palatable (E. T. Baconawa, 1986).

We will now discuss the nutritional considerations of feeding sugarcane and by-products, the upgrading of the by-products of sugarcane tops and bagasse and the practice of feeding sugarcane in one country, the Philippines.

The Nutritional Considerations of Feeding Sugarcane and By-products.

This topic has been previously covered by Leng and Preston (1985) and Leng and Preston (1986). In their discussion they highlight attention to the following: 1) optimize ruminal digestion by obtaining a high microbial protein relative to energy ratio, high fibre digestibility for energy and high propionate or glucogenic potential relative to acetate and butyrate and 2) supplementation through manipulation of diet by providing bypass protein, bypass starch to provide extra glucose and fat (R. A. Leng and T. R. Preston, 1986).

The efficiency of microbial cell synthesis or yield (the Y ATP) in g microbial cells per mole of ATP from fermentation of energy substrate is influenced by the content in fermentable nitrogen (N) used for synthesis. This N can be in the form of urea N, chicken litter or high protein forages; about 30 g N/kg of fermentable carbohydrate is required; urea N has been found to be a more efficient source of fermentable N than poultry litter and poultry litter can add further to the effect of urea N; peptides and amino acids from relatively insoluble protein can stimulate microbial growth (R. A. Leng and T. R. Preston, 1986). The Y-ATP is also positively correlated with rumen outflow rates with greater rumen volume or distension due to slowly digested fibre; it has been suggested that 8 g/kg of body weight be provided as 'good quality' forage that slowly breaks down and provides for good rumen outflow and encourages good intake of molasses-based diets; legumes are considered as ideal sources for this (R. A. Leng and T. R. Preston, 1986). Protozoa can also affect Y ATP of rumen microbes; because protozoa engulf bacteria and are well-established because they assimilate soluble sugars, defaunation or their removal, for example by chemical means, will result in greater flow of microbial protein (R. A. Leng and T. R. Preston, 1986).

Fishmeal is a source for bypass protein and oils, rice polishings provide glucose from bypass starch, bypass protein and fats and maize grain supplies glucose from bypass starch and oil (R. A. Leng and T. R. Preston, 1986).

There is a need for both bypass protein and fat for growth; increasing the proportion of glucogenic energy or propionate increases nitrogen balance of young lambs; as well, the proportion of propionic acid in rumen VFA increases the efficiency of use of metabolizable energy for fattening in sheep; with lactating animals there is a need for protein, fat and glucose necessary for lactose and glycerol (the fatty acid backbone) synthesis and oxidation of glucose for NADPH for the synthesis of acetate to C16-18 long chain fatty acids (R. A. Leng and T. R. Preston, 1986).

For diets with sugarcane tops it has been suggested to treat the fibre with alkali to increase its digestibility, add urea at 2% dry matter (DM) to provide more fermentable N, supplement with oil seed cake containing bypass protein and dietary lipid, increase glucogenic potential by manipulating propionate to VFA ratios with monensin, feeding bypass starch from rice and maize grain, add 'good quality' forage which is related to supporting good intake with green legumes or young grasses and supplementing all minerals particularly sulfur (S) (R. A. Leng and T. R. Preston, 1985, T. R. Preston, 1986).

Bagasse is high in lignified (low digestible) fibre, low in fermentable N, protein and long chain fatty acids (R. A. Leng and T. R. Preston, 1985). Treatment can be with alkali or steam (R. A. Leng and T. R. Preston, 1985). Steam treatment with decompression has recently been investigated as a method for separating out the hemicellulose, cellulose and lignin components of fibre (Orskov, 2000). Steam treated basal diets of bagasse can be fed with fishmeal (protein and fat) and maize (bypass starch and fat) (R. A. Leng and T. R. Preston, 1985). Treated bagasse can be supplemented with urea N and molasses as low-cost substitute with cereal straws (D. V. Rangnekar, 1988). Untreated bagasse can be supplemented with green legumes, urea N, molasses along with rice polishings, cottonseed meal or copra meal (D. V. Rangnekar, 1988).

Upgrading Sugarcane By-products of Sugarcane Tops and Bagasse.

Sugarcane tops and bagasse are the major by-products of milling the sugarcane plant.

Sugarcane tops which is characterized by its low digestible fibre, low sugar content (R. A. Leng and T. R. Preston, 1985) and what N there is would make it amenable to new biotechnological approaches proposed to upgrade this by-pro-duct feed.

The recent proposal to further elucidate microbial mechanisms in anaerobic ligninolysis to be used with ensilage of straw can also be proposed to improve the digestibility of sugarcane tops. Progress towards this type of research has for example been research with mesophilic digestor fed waste-activated sludge which has been observed to lead to fragmentation of oligolignols including single rings and smaller compounds (P. J. Colberg and L. Y. Young, 1982). Also, the attack on lignocellulose has been observed in the rumen with rumen isolate *strain 7-1* (D. E. Akin, 1980). Research should progress to further study and isolate intermediates in lignin breakdown useful in describing the organic chemical basis indicating the associated enzymatic activities of lignin breakdown and the genetic basis of microbial metabolism established and engineered into microbial hosts in ensilage.

There is another approach to improve the digestibility of fibre in sugarcane tops which is to reduce lignin content by genetic engineering. This involves down-regulation of lignin biosynthesis with anti-sense repression which would use anti-sense-oriented mRNA (the opposite strand of mRNA that is transcribed) to deactivate mRNA. There are also approaches to genetically mutate various biosynthetic activities. One approach involving the down-regulation of O-methyl transferase activity in poplar plants resulted in 70% reduction in CCoAOMT activity associated with a 60% decrease in lignin (R. Zhong et al. 2000). The down-regulation of lignin applies to the whole plant and would reduce lignin content in bagasse.

There has been a recent development in the U.K. where high-sugar (high fructan) ryegrass varieties were developed containing as much as 50% more sugar (Anon., 2003a, Anon., 2003b and S. Young, 2004). Study has begun on genes which control sugar metabolism (Anon., 2004). The understanding of the regulation may lead to the cloning of fructan metabolism genes in sugarcane including the sugarcane tops. Fructans are in part what are called water-soluble carbohydrates (WSC). WSC have been implicated in limiting microbial protein synthesis in the rumen. There is higher intake likely from greater digestibility and a better balance between organic matter (OM) digested or energy supply and the N degraded for microbial protein synthesis and greater microbial N flow and leads to a higher proportion of protein incorporated into milk protein from dietary protein on a high WSC diet (S. Young, 2004).

Although the N in sugarcane tops may be supplemented, it might be of further help to conserve needed micronutrients of peptides and amino acids that are there. It has been hypothesized, as was already referred to, that slowing down the break-down of protein and the release of peptides and amino acids within the rumen stimulates microbial growth by making them more available (R. A. Leng and T. R. Preston, 1986). It has been found that plant proteases contribute to the initial stages of proteolysis of grazed herbage (W.-Y. Zhu et al., 1999). Down-regulation of proteases could help conserve plant protein. Pre-wilting forages also helps protect protein. To what extent sugarcane tops pre-wilted prior to ensiling need to be down-regulated for protease activity remains to be seen.

An area for further research of sugarcane is ensiling of sugarcane tops after the milling season, as has been referred to. This is ideally when the cane is mature and of greater feeding value due to its higher sugarcane content and fed year round. The use of ammonia or ammonia in situ with urea has been used to preserve the mass which ensures development of lactic acid-producing bacteria rather than yeasts, which convert sugar to alcohol (T. R. Preston, 1986). It has also been proposed that anhydrous ammonia be used to preserve sugarcane tops and which would preserve the sugars (R. A. Leng and T. R. Preston, 1985). The addition of molasses can supplement what sugar there is and helps to preserve the ensiling mass (R. A. Leng and T. R. Preston, 1985). The sugars would improve microbial growth in the rumen (R. A. Leng and T. R. Preston, 1985). The proposal to genetically boost WSC such as fructans may serve to make sugars more available for preservation with ensilage as well as stimulate microbial growth. Sugarcane tops can also be ensiled with better quality crops together with molasses (D. V. Rangnekar, 1988).

The bagasse can be treated chemically and physically (R. A. Leng and T. R. Preston, 1985), as was mentioned. A more potent alkali is anhydrous ammonia or ammonium hydroxide used with cereal straws with pH reaching 10 or above with the solubilization of hemicellulose and breakdown of lignin bonds (P. T. Doyle et al., 1986). Urea pretreatment or treatment on the other hand, which produces ammonium carbonate and bicarbonate, may only reach pHs of between 8-9 which would not break down lignin bonds as effectively (P. T. Doyle et al., 1986). Steam treatment has also been used successfully on bagasse to improve its palatability and digestibility (R. A. Leng and T. R. Preston, 1985, D. V. Rangnekar, 1988). Steam treatment has recently been investigated to separate out the hemicellulose, cellulose and lignin components of fibre (Orskov, 2000).

Bagasse might also be processed with what is termed as exogenous fibrolytic enzymes (EFEs), including enzymes and yeasts as was referred to in a paper by Bocanawa (1986) in the Philippines. EFEs include preparations from yeast (Y. Wang and T. A. McAllister, 2002). Enzymes include xylanases and cellulases as well as other secondary activities (Y. Wang and T. A. McAllister, 2002). There is a need to improve the approach in determining the fibrolytic activities and their sources, to determine the different methods of application of EFEs (eg. spraying to stabilize enzymes and optimizing prehydrolysis treatment time) and to study the effect of moisture content of the feed (Y. Wang and T. A. McAllister, 2002) when utilizing EFEs and the use of other methods to aid EFEs to breakdown lignin such as the pretreatments with alkali (Y. Wang and T. A. McAllister, 2002) and use of lignases such as lacasses with low mass mediators, used in process-scale bio-bleaching of pulp, with NO, NOH and HRNOH groups termed the Lignozym ® process (D. O. Krause et al., 2003) and esterases (egs. acetylesterase and ferulic acid esterase) and etherases which breakdown cross bridges or linkages (Y. Wang and T. A. McAllister, 2002, D. O. Krause et al., 2003).

Solid state fermentation (SSF) with aerobic fungal culture may be also applied to the treatment of bagasse. The SSF process which is termed as such because of the fungi requiring low water activity on the substrate (R. L. Howard et al., 2003) involves white rot basidiomycetes (F. Zadrazil et al., 1995). However, there is a need to boost the levels of enzymes and their Kcat for sufficient biological pretreatment (D. G. Armstrong and H. J. Gilbert, 1991). Regulation of lignases should be constitutive rather than under secondary metabolic control. There might be a need to engineer increased pH and thermostability in the future (R. L. Howard et al., 2003). There is a need to decrease incubation times to avoid sizable losses of organic matter (OM) (F. Zadrazil et al., 1995). Cel (cellulase) negative mutants have also been used with this in mind (F. Zadrazil et al., 1995). *Coprinus, Cyathus stercoreus* and *Dichomitus squalens* are examples of fungi that have been used (P. T. Doyle et al., 1986, F. Zadrazil et al., 1995). The use of SSF with aerobic fungi requires a degree of control and specialist monitoring making the technology prohibitive for small farmers (F. Zadrazil et al., 1995). Expensive, large-scale commercial units that may be developed and accesssing them present problems (R. M. Acharya, 1988, F. Zadrazil et al., 1995).

The use of EFEs fed with sugarcane tops has yet to be proposed as there are no reports found in the literature and so with the application of the SSF process. The use of anaerobic ligninolysis with ensilage or the treatment of bagasse can also be proposed with future development.

The Practice of Feeding Sugarcane in Countries Like the Philippines.

In a country that traditionally grows sugarcane for sugar production replacing natural growing grasses especially when they are scarce with sugarcane presents a possible breakthrough for sugarcane production in the Philippines (E. T. Bocanawa, 1986).

In 1985 production of sugarcane was reported annually at 23 million metric tons of which 4.70 million metric tons was sugarcane tops, 6.45

million metric tons was bagasse and 916,000 metric tons was molasses (E. T. Baconawa, 1986).

The practice of growing sugarcane would involve harvesting the sugarcane every 10-12 months and can be ensiled for feeding (E. T. Baconawa, 1986).

Larger commercial operations exist in the Philippines including farms that plant and feed whole sugarcane and sugarcane estates that plant and mill sugarcane and feed them to cattle (feeders) and dairy (E. T. Baconawa, 1986). In one sugarcane plantation, 60% whole chopped sugarcane was used together with 40% chopped napier grass fed to dairy cows at 15-20 kg/day with 10-15 kg fresh brewer's grains; average daily production was 8.5 kg of milk/day during a period of 260 days; some selected cows produced 15 liters/day (E. T. Baconawa, 1986). Sugarcane tops are also collected from fields during the milling season by small farmers; medium and large livestock raisers also feed sugarcane tops chopped and fed or ensiled during the off-milling season; sugarcane tops are used in a semi-commercial and commercial scale where sugarcane is widely grown where sugar planters use it for feeder and breeder cattle and water buffaloes (E. T. Baconawa, 1986).

Sugarcane bagasse is another by-product and a potential animal feed. The need for its treatment makes it an expensive proposition because of the cost of chemicals. The use of enzyme or yeast is promising. The fermentation of bagasse has been proposed for the Philippines (E. T. Baconawa, 1986). As has been previously mentioned, use of EFEs or enzymes and other methods to aid the break down of lignin and SSF are promising technologies for use with bagasse.

Steam treatment which has recently been investigated with cereal straw in separating out components of hemicellulose, cellulose and lignin and is promising (E. R. Orskov, 2002) needs further attention in the Philippines.

Molasses is also produced and recommended for feeding at 4-5% for various livestock species and can be given as a lick in the feedlot. Molasses can be used in the summer when forage or pasture is scarce when it is fed with rice straw at 8-10% which adds to its palatability (E. T. Baconawa, 1986).

The Practice of Feeding Sugarcane in Countries Like Brazil and Australia.

In some countries with ample arable land area, range farming is practiced with feedlots for cattle feed a variety of forages both grown from pasture and plantations. In Brazil a case can be made for both range and plantation farming using similar operations with high-energy, high-protein forages and grains with new proposals for sugarcane and its by-product cropping.

Plantations can readily supply SCT that is available upon harvest and that can be field-dried and transported to feedlots in commercial plantations and surrounding commercial feedlots and then ensiled with high performance qualities such as boosted WSC content, high-protein (with amino acid supplementation such as lysine, cysteine and methionine) and low-lignin in rolled yeast fermented bagasse from mills with GM high-protein/amino acid (e.g. threonine) grain.

Short Summary.

One advantage of the sugarcane plant as a crop is it being an efficient biomass producer. The integration of livestock production with sugarcane production has been encouraged in countries like the Philippines. Dairy production has produced about 8.5 kg of milk / day during a lactation period of 260 days while 0.84 kg / day ADGs have been attained with a diet of sugarcane tops and rice polishings. It has been proposed that bagasse be used with treatment or fermentation in the Philip-pines. Molasses can also be used as a feed at a 4-5% recommended rate and at 8-10% addition with rice straw to improve palatability. To feed sugarcane, the fol-lowing considerations are highlighted: optimizing ruminal digestion by obtaining a high microbial protein relative to energy ratio, high fibre digestibility for energy and high propionate or glucogenic potential and supplementation by providing bypass protein, bypass starch for glucose and fat. The Y ATP in g microbial cells per mole of ATP is influenced by fermentatable N in the form of urea N, chicken litter or high protein forages; about 30 g N/kg of fermentable carbohydrate is required; urea N has been found to be a more efficient source of fermentable N than poultry litter; peptides and amino acids from relatively insoluble protein may stimulate microbial growth. The Y ATP is also positively correlated with rumen outflow rates with greater rumen distention due to slowly digested fibre with 8g/kg body weight suggested as 'good quality' forage that encourages good rumen outflow and intake on molasses-based diets; legumes are ideal sources. Protozoa can also affect the Y ATP, because protozoa engulf bacteria, with defaunation or removal by chemical means resulting in greater flow of microbial protein. Fishmeal provides bypass protein and oils, while rice polishings is a sources for bypass starch, bypass protein and fats and maize grain supplies glucose from bypass starch and oil. There is a need for both bypass protein and fat for growth; increasing the proportion of glucogenic energy or propionate increases nitrogen retention of young lambs; as well increasing the proportion of propionic acid increases the efficiency of use of metabolizable energy for fattening in sheep; with lactating animals there is a need for protein, fat and glucose necessary for lactose and glycerol (fatty acid backbone) synthesis and oxidation to NADPH for synthesis of acetate to C16-18 long chain fatty acids. Sugarcane tops should be treated with alkali, urea added at 2% DM to provide fermentable N, supplemented with oil seed cake for bypass protein and lipid, the ratio of propionate to other VFAs manipulated with monensin, fed with rice and maize grain for starch and 'good quality' forage to support good intake such as green legumes and grasses and supplementing all minerals. Bagasse can be treated with alkali or steam. Treated bagasse supplemented with urea N and molasses can be a substitute for cereal straw and untreated bagasse can be fed with green legumes, urea N, molasses along with rice polishings, cottonseed meal or copra meal. New biotechnological approaches are proposed for the sugarcane plant to improve the nutrient value of by-products including the sugarcane tops and bagasse. Microbial anaerobic ligninolysis with ensilage of sugarcane tops is proposed. Progress towards this type of research has been for example with mesophilic digestor fed waste-activated sludge and with rumen isolate strain 7-1. Microbial mechanisms need to be further elucidated and engineered into microbial hosts in ensilage. Reducing lignin content by genetic down-regulation such as with anti-sense repression would improve the digestibility of both tops and bagasse. In one study, down-regulating CCoAOMT activity successfully reduced lignin content. The development of high-sugar ryegrasses with the study of genes that control fructan metabolism might lead to the cloning of fructan metabolism genes in sugarcane including sugarcane tops. Fructans, in part termed as WSC, may limit microbial protein synthesis as has been pointed out. There is improved intake likely due to improved digestibility and improved microbial protein flow from a better balance between OM digested and N degraded with greater production of milk protein on a high WSC diet. It has also been proposed to down-regulate protease activity in sugarcane to protect protein making peptides and amino acids more available; pre-wilting of forage may also do so further to this and this still remains to be seen. An area that needs further research is the ensilage of sugarcane tops. This includes the use of ammonia or ammonia in situ from urea which helps ensure development of lactic acid bacteria rather than yeasts which convert sugars to alcohol. Anhydrous ammonia has been proposed to preserve the ensiling mass to preserve the WSC or the addition of molasses to provide sugars for preservation. Both would provide sugars to stimulate microbial growth in the rumen. Genetically boosting fructan content could do the same. Bagasse can be processed chemically and physically. A potent alkali is anhydrous ammonia or ammonium hydroxide with a pH of 10 or above compared with in situ urea where only a pH of between 8-9 is attained; steam treatment of bagasse improves its palatability and digestibility. EFEs may be used with bagasse although variables exist as to its optimal use including the use of other methods to aid fibrolysis to break down lignin; alkali, lacasses with low mass mediators, esterases and etherases are proposed. SSF requires further development including the need to boost enzyme levels, increase the Kcat and make production of lignases constitutive; there is also a need to avoid OM losses by decreasing incubation times and the use of cel negative mutants. The application of SSF is still prohibitive. EFEs and the SSF process can also be applied to feeding sugarcane tops and so with anaerobic ligninolysis with ensilage or treatment of bagasse. In the Philippines which grows sugarcane for sugar production, replacement of grasses with sugarcane poses a possible breakthrough. In 1985, annual production was reported at 23 million metric tons of sugarcane of which 4.70 million metric tons was sugarcane tops, 6.45 million metric tons was bagasse and 916,000 metric tons was molasses. Commercial operations either in farms or plantations, that plant and mill sugarcane, in one farm 60% chopped sugarcane and 40% napier grass at 15-20 kg / day with 10-15 kg fresh brewer's grain was fed with average production of 8.5 kg milk/ day with selected cows producing 15 liters/ day; small farmers also collect sugarcane tops during milling season with medium to large livestock raisers feeding chopped sugarcane tops or ensiling it during the off-milling season; sugarcane tops are also fed in a semi-commercial and commercial scale in a sugarcane growing area for feeders and breeder cattle and water buffaloes. In the Philippines, the cost of treating bagasse with chemicals is expensive; the use of enzyme and yeast is promising. Fermentation of bagasse has been proposed for the Philippines. Steam treatment which separates out the components of cellulose, hemicellulose and lignases also needs further attention. Finally, molasses can be fed at 4-5% for various livestock or as a lick and can be fed with rice straw at 8-10% to improve palatability when forage or pasture is scarce. Range farming with forages including use with sugarcane with feedlot feeding with plantations is a new proposal including in countries like Brazil and Australia and with sugar can tops transported for surrounding feedlot farms using GM, high-performing WSC-containing, high-protein (including in essential amino acids such as lysine, methionine and cysteine) and low-lignin with sugarcane tops used with complete rolled yeast fermented bagasse feed with high-energy and high-protein/amino acids from grain.

References.

1. R. M. Acharya. 1988. Keynote address. In: Non-conventional Feed Resources and Fibrous Agricultural Residues. C. Devendra (Ed.). IDRC-ICAR Ottawa Canada.
2. D. E. Akin. 1980. Attack on lignified grass cell walls by a facultative anaerobic bacterium. Appl. Env. Microb. 40: 809-820.
3. Anon. 2003a. Integrated resesearch leads to award-winning grass and oat varieties. business October issue: 5-6.
4. Anon. 2003b. Success tastes sweet for high-sugar grasses. http:// www.igerhbsrc. ac.uk/New/PR/2003/22April03successtastessweetfo rhighsugargrasses.
5. Anon. 2004. V. J. N. Nagaraj. http://www.unibas.ch/bothebel/ people/vinay/ vinay.htm.
6. D. G. Armstrong and H. J. Gilbert. 1991. The application of biotechnologies for future livestock production. In: Physiological Aspects of Digestion and Metabolism in Ruminants. T. Tsuda, Y. Sasaki and R. Kawashima (Eds.). Pp. 737-761. Academic Press. San Diego USA.
7. E. T. Baconawa. 1986. Case study-Prospects for reconversion of sugarcane into animal feed in the Philippines. In: Sugarcane as Feed. R. Sansoucy, G. Aarts and T. R. Preston (Eds.). FAO Animal Production and Health Paper 72. FAO-UN Rome, Italy.
8. P. J. Colberg and L. Y. Young. 1982. Biodegradation of lignin-derived molecules under anaerobic conditions. Can. J. Microbiol. 28: 886-889.
9. P. T. Doyle, C. Devendra and G. R. Pearce. 1986. Improving the feeding value through pretreatments. In: Rice Straw As A Feed for Ruminants. Pp. 54-89. IDP Canberra Australia.
10. R. L. Howard, E. Abotsi, E. L. Jansen van Rensburg and S. Howard. 2003. Lignocellulose biotechnology: issues of bioconversion and enzyme production. African J. of Biotechnology 2: 602-619.
11. D. O. Krause, S. E. Denman, R. I. MacKie, M. Morrison, A. L. Rae, G. T. Attwood and C. S. McSweeney. 2003. Opportunities to improve fibre degradation in the rumen: microbiology, ecology and genomics. FEMS Microbiol. Rev. 27: 663-693.
12. R. A. Leng and T. R. Preston. 1985. Constraints to the efficient utilization of sugarcane and its by-products as diets for production of large ruminants. In: Ruminant Feeding Systems Utilizing Fibrous Agricultural Residues-1985. R. M. Dixon (Ed.). Pp. 27-47. IDP Canberra Australia.

13. R. A. Leng and T. R. Preston. 1986. Constraints to the efficient utilization of sugarcane and its by-products as diets for production of large ruminants. In: Sugarcane as Feed. R. Sansoucy, G. Aarts and T. R. Preston (Eds.). FAO Animal Production and Health Paper 72. FAO-UN Rome, Italy.

14. M. R. Naseevan. 1986. Sugarcane tops as animal feed. In: Sugarcane as Feed. R. Sansoucy, G. Aarts and T. R. Preston (Eds.). FAO Animal Production and Health Paper 72. FAO-UN Rome, Italy.

15. E. R. Orskov. 2002. Crop Fractionation. In: Trails and Trials in Livestock Re-search. Pp. 67-69. International Feed Resources Unit, Aberdeen, U. K.

16. T. R. Preston. 1986. Sugarcane as animal feed: an overview. In: Sugarcane as Feed. R. Sansoucy, G. Aarts and T. R. Preston (Eds.). FAO Animal Production and Health Paper 72. FAO-UN Rome, Italy.

17. D. V. Rangnekar. 1988. Availability and intensive utilization of sugarcane by- products. In: Non-conventional Feed Resources and Fibrous Agricultural Resi- dues. Strategies for Expanded Utilization. C. Devendra (Ed.). Pp. 76-93. IDRC ICAR Ottawa Canada.

18. Y. Wang and T. A. McAllister. 2002. Rumen microbes, enzymes and feed digestion (Internet document).

19. S. Young. 2004. Sweet grass, contented cows. http://www.cherrybyte. org/Articles/Agriculture/CowsEating/Cowsea . . .

20. F. Zadrazil, A. K. Puniya and K. Singh. 1995. Biological upgrading of feed and feed components. In: Biotechnology in Animal Feeds and Animal Feeding. R. J. Wallace and A. Chesson (Eds.). Pp. 50-70. VCH Weinheim FRG.

21. R. Zhong, W. H. Morrison III, D. S. Himmelsbach, F. L. Poole II and Z.-H. Ye. 2000. Essential role of caffeoyl coenzyme A O-methyltransferase in lignin biosynthesis in woody poplar plants. Plant Physiology 124: 563-578.

22. W.-Y. Zhu, A. H. Kingston-Smith, D. Troncoso, R. J. Merry, D. R. Davies, G. Pichard, H. Thomas and M. K. Theodorou. 1999. Evidence of a role of plant proteases in the degradation of herbage proteins in the rumen of grazing cattle. J. Dairy Sci. 82: 2651-2658.

Chapter 8

FEED RESOURCES FOR RUMINANTS IN ASIA AND POSSIBLE ON-FARM TECHNOLOGIES TO IMPROVE UTILIZATION.

The Problem and the Potential.

Small-holder farmers constitute the majority of the farming population in the developing countries in Asia (D. B. Roxas et al., 1997).

Ruminant production systems in Asia are unlikely to change in the foreseeable future although shifts towards intensification are also envisaged (C. Devendra and C. C. Sevilla, 2002).

Feed resources and nutrition are the principal constraints affecting animal production (C. Devendra and C. C. Sevilla, 2002).

We will discuss here types of feed resources available to farmers in countries with reference to Asia. These are grasses, legumes and other fodders, crop residues, some of which could be agro-industrial by-products (AIBPs), non-conventional feed resources (NCFRs) and fodder trees and shrubs (C. Devendra, 1997, C. Devendra and C. C. Sevilla, 2002).

There is inadequate adaptive research that promotes wide adoption and use of appropriate technologies (C. Devendra and C. C. Sevilla,

2002). We will include here discussion on the widely researched chemical pre-treatment of feed residues, use of supplementation with feeds, including the use of multi-nutrient block licks, the recent development of high-sugar ryegrass varieties in the U. K., which have resulted in increased efficiency of microbial protein synthesis in the rumen (Anon., Anon., 2003), with grass varieties in coconut lands in the tropics (Anon., O. O. Parawan and H. B. Ovalo, 1986) cloned with fructan (a sugar or water-soluble carbohydrate (WSC)) metabolism genes, the proposed use of varieties that have been selected for improved straw quality (D. B. Roxas et al., 1997, C. Devendra, 1997, C. Devendra and C. C. Sevilla, 2002), protein enrichment, the issue of protozoa in the rumen, which decreases the protein to energy ratio, when defaunated, improves productivity, limited by amino acid supply, such as body weight gain, wool growth and milk production (R. A. Leng, 1992) and their forages with anti-protozoal agents, and further, their genetic engineering in forages, anti-nutritional properties and problems of toxicity in fodder trees and shrubs due to tannin content (C. Devendra and C. C. Sevilla, 2002) controlled with the use of genetic engineering and finally down-regulating protease activity in forage.

It has been noted that although pre-treatment and supplementation have most dominated research there still has been no discernable impact on small farms (C. Devendra, 1997). Included are physical treatment (i.e. steam pressure treatment), urea-ammonia pre-treatment of cereal straws, use of AIBPs and NCFRs, use of forages as secondary crops to food crops and the use of fodder trees and shrubs. Various biotechnologies can be used to address the need to improve fibre availa-bility including the engineering of low-lignin forages coupled with urea pre-treatment which might present a synergy in making fibre more available. There are the proposed use of anaerobic ligninolysis with ensilage or fermentation and fungal solid-state fermentation, use of enzymes (e.g. etherases, esterases) and the more advanced use of lacasses in reactors to help breakdown and bleach lignin in low-lignin feed residues such as straw.

Also, not included further in this discussion, is the recent proposed use of somatic embryogenesis to increase fibre production in cereal straw and bagasse with, for e.g., improved standability in low-lignin, dual-purpose cereal straws.

Finally, discussed are the feed resources in one country, the Philippines, taken from F. A. Moog (2002).

Grasses, Legumes and Other Fodders.

Poor quality native grasses together with fibrous residues of a wide range of crops form the principal feeds for ruminants on small mixed farms (C. Devendra and C. C. Sevilla, 2002).

Grasses are available from native grasslands, rangelands, forests, fallows, wastelands, roadsides and cultivated areas after crop harvest; livestock can be free-grazing, tethered or feed cut-and-carried when they are confined (C. Devendra and C. C. Sevilla, 2002).

Pastures in Thailand are improved with communal grazing lands and roadsides oversown with *Stylosanthes hamata*; for backyard pastures *Brachiara ruziziensis, Panicum maximum* cv Hamil, *Panicum maximum* and *Pinneseteum purpureum* are used (D. B. Roxas et al., 1997).

In Southeast Asia there is an estimated 210 million hectares (ha) for pasture improvement with perennial tree crop systems with combinations of cattle under coconut, oil palm and mango, sheep under coconut, rubber and durian and goats under coconut; most attention has been given to integration of cattle with coconut where light penetration benefits the understory (C. Devendra and C. C. Sevilla, 2002). In Malaysian rubber and oil palm estates and along hedgerows in Indonesia grasses such as *Setaria splendida* and *Tripsacum laxum* to control soil erosion and hybrid napier or *P. purpureum* x *P. glaucum*, with higher yields, have been used (D. B. Roxas et al., 1997).

Typical grass weeds associated with rice are *Echinochloa crusgalli, E. colonum, E. cruspavonis* and *Leptochloa chinensis* (C. Devendra and C. C. Sevilla, 2002).

Improved forages, in particular legumes, can be used with food crops as inter- and relaycrops, sequence crops in rotations and improvement of fallows (C. Devendra and C. C. Sevilla, 2002). Shorter duration crops and new or improved irrigation schemes have contributed to intercropping (C. Devendra, 1997). Food-feed intercropping is viewed not only as a contribution to human food supplies in the short term but also for its contribution to animal feeds and its sustainability over the long term (C. Devendra, 1997).

In India, a rice-wheat-cowpea cropping sequence yields grain, straw and green fodder (C. Devendra and C. C. Sevilla, 2002). Oats, berseem and lucerne was planted after rice harvest resulted in dry matter yield (C. Devendra and C. C. Sevilla, 2002). Sun hemp intercropped with mung bean and pigeon pea produce grain and forage (C. Devendra and C. C. Sevilla, 2002)

In the Philippines rice-mung bean and rice-mung bean/siratro yielded fodder from mung bean and siratro including additional clippings for siratro as green manure for the following rice crop; rice produced grain and straw (C. Devendra, 1997, C. Devendra and C. C. Sevilla, 2002). The rice-mung bean+siratro cropping pattern produced, in 1000 square meters of land, forage to support 1 cow for 4 months in the dry season (C. Devendra and C. C. Sevilla, 2002).

In Pakistan cowpea was intercropped with maize producing more crude protein and dry matter (C. Devendra and C. C. Sevilla, 2002).

In the Philippines, *Sesbania* and *Desmanthus* have increased rice yield and produced forage (D. B. Roxas et al., 1997) and cow pea and pigeon pea have also been used with rice (C. Devendra, 1997).

In South Asia lentils, chickpea, groundnut and lathyrus have been used with rice (C. Devendra, 1997).

In India's four northern states 30% of the cropped area irrigated farming systems with small holders have been sown with forage crops; in Pakistan, in Punjab province, 15% have been sown with forage crops (C. Devendra and C. C. Sevilla, 2002). Fodders sown in the summer include pearl millet, maize, sorghum and cowpea and in the winter berseem, lucerne, rapeseed and oats; in Bangladesh the legume *Lathyrus sativus* has been integrated with rice cropping systems on small mixed farms for dairy cattle (C. Devendra and C. C. Sevilla, 2002).

In South Asia in rain-fed production systems there are large fallows that could be used to establish leguminous forages without interfering with crop production (C. Devendra and C. C. Sevilla, 2002).

Legumes can be used as cover-crops in perennial tree crop plantations; in rubber and oil palm plantations, leguminous crops have been planted to control weeds and contribute to early growth of trees through nitrogen accretion (C. Devendra and C. C. Sevilla, 2002).

Low-Quality Residues.

Low-quality residues refer to various by-products from crops and are a major group of residues having low crude protein of 3-4% and high crude fibre content of 35-48% and include cereal straws, cereal stalks, cereal stovers, legume haulms, sugarcane tops, bagasse, cocoa pod husks, pineapple waste, coffee seed pulp and palm press fibre and form the base of feeding systems for ruminants throughout the developing countries (C. Devendra, 1997, C. Devendra and C. C. Sevilla, 2002). They provide for maintenance and some production of ruminants (C. Devendra, 1997). Cereal straws account for about 94% of the total supply of fibrous crops residues (FCRs), a term synonymous with low-quality residues, in Asia (C. De-vendra, 1997).

For rice straw, calculating the requirement of buffaloes and cattle given their weight and intake, availability is well over the requirement thus reflecting con-siderable underutilization by ruminants (C. Devendra, 1997).

Medium-Quality Residues.

Crop residues, also including AIPBs, that have medium protein content including coconut cake, palm kernel cake and sweet potato vines are referred to as medium-quality residues (C. Devendra, 1997).

Good-Quality Residues.

Good-quality residues, including again AIBPs, high in protein, high in energy and minerals include oil seed cakes and meals such as gound nut cake, soyabean meal and cassava leaf meal (C. Devendra 1997, C. Devendra and C. C. Sevilla, 2002).

Non-conventional Feed Resources (NCFRs).

These feeds are defined as those not used traditionally in animal feeding and are also a major group include those such as cocoa pod husks, rubber seed meal, distiller solubles, shrimp waste, leather shavings and poultry litter (C. Devendra and C. C. Sevilla, 2002).

Fodder Trees and Shrubs.

Despite a wide diversity, research and development has tended to be focused on a narrow range of species with *Leucaena leucocephala* and *Gliricidia sepium* in Asia and Africa and *Erythrina spp.* in Latin America (C. Devendra, 1997). *Leucaena leucocephala* used most widely in Asia has resulted in significant increases in live weight gain or milk yield and reduced cost as a result of feeding because feeding reduces use of concentrates (C. Devendra, 1997, C. Devendra and C. C. Sevilla, 2002).

It has been estimated that if agricultural lands in Asia were planted with 230 fodder trees and shrubs per hectare, 150 extra tons of fodder per ha could be produced annually; this could support a total of 56 million animal units, given as the sole diet, during the dry season (C. Devendra and C. C. Sevilla, 2002).

The three-strata forage system (TSFS) developed in Indonesia and also used in other countries such as India and sub-Saharan Africa is planted in 0.25 ha and provides pasture in the wet season, shrubs in the mid dry season and trees in the late dry season with cash crops planted in the centre (D. B. Roxas et al., 1997, C. Devendra and C. C. Sevilla, 2002). Benefits from the TSFS are increased forage production, higher stocking rates and liveweight gains, increased farm income, reduced soil erosion, increased soil fertility and a supply of fuel wood (D. B. Roxas et al., 1997, C. Devendra and C. C. Sevilla, 2002).

Another scheme of utilizing fodder trees and shrubs in a form of upland farming called sloping agricultural land technology (SALT) utilizes alleys of 4-5 meter wide sown with annual and perennial crops between contoured double hedgerows of leguminous fodder trees and shrubs including *Calliandria calothysus, Leucaena leucocephala, Leucaena diversifolia, Gliricidia sepium* and *Fleminga macrophylla*; hedgerows can provide for mulch and green manure; every third row is planted with perennial crops while others are planted with cereals, grain legumes and other annual crops (C. Devendra and C. C. Sevilla, 2002). There are variants to SALT: SALT-2 which is simple agro-livestock technology, SALT-3 which is sustainable agro-forestry land technology and SALT-4 which is small agro-fruit livelihood technology (C. Devendra and C. C. Sevilla, 2002). SALT-2 is a goat-based agro-forestry project comprising land use of 40% for agriculture, 40% for livestock and 20% for forestry; there is a net profit of USD\$ 213 per month per hectare (C. Devendra and C. C. Sevilla, 2002). In the Philippines 24 million ha of sloping uplands are planned for development; use has been extended to India, Sri Lanka and Laos (C. Devendra and C. C. Sevilla, 2002).

Chemical Pre-treatment of Cereal Straws.

Among treatment of cereal straws with alkali, urea-ammonia pre-treatment has been the most significant one (C. Devendra, 1997). Variables in applying this pre- treatment include level of urea, whether the straw is long or chopped form, application by spraying or impregnation, moisture content of straw and storage in open and closed containers (C. Devendra, 1997).

One study in Sri Lanka involved giving upgraded rice straw with urea and rice straw supplemented with urea. Upgrading consisted of impregnating a solution of urea with storage in sealed polyethylene bags while supplementation consisted of adding urea to the straw. Growth rates were 718 compared with 217 g per head per day with higher intakes of 2.4% compared to 1.8% of liveweight (C. Devendra, 1997).

There are constraints that have prevented small-scale farms from adopting this technology; economic implications depend on comparing the cost of inputs with the price of beef (D. B. Roxas et al., 1997 and C. Devendra, 1997).

Sodium hydroxide among the hydroxides has also received a great deal of attention with wet and "dry" processes used although there is more of a hazard in handling it and without nitrogen enhancement of the straw (C. Devendra and C. C. Sevilla, 2002).

Supplementation with Feeds.

Supplementation can help promote efficient microbial growth in the rumen as well as increase the non-degraded portion in the small intestines (C. Devendra, 1997).

Supplements have already been mentioned in the previous sections on feeds resources other than low quality residues. These include grasses and legumes and other fodders with grasses from various areas including grazing lands and roadsides and backyard pastures, pasture improved with perennial tree crop systems, grass weeds, improved forages, in particular legumes, used with food crops in inter- and relaycrops, sequence crops in rotations and improvement of fallows, which are forms of producing forage crops as a secondary crop to food crops, and as cover-crops in tree crop plantations, cakes and meals or AIBPs, such fodders as cassava leaf meal and sweet potato vines, NCFRs, in the TSFS with cash crops, grasses and herbaceous legumes, shrub legumes and fodder trees and in SALT alley cropping with perennial and annual food crops, fodder trees and shrubs.

A successful example of supplementation involved the use of cotton seed cake (CSC) in Henan province in China for beef production on a basal diet of urea-treated straw; profit increased with use of urea-treated straw and increased ten-fold with CSC supplementation; beef production with this local feed is viable for small farms who use it with good market access (C. Devendra and C. C. Sevilla, 2002).

Forage production in small-holder farms is restricted mostly to irrigated systems and mostly observed in South Asia; in rain-fed production systems of Asia the adoption of food crop-feed systems on small farms is still very low; often the inclusion of forages as a secondary crop in food crops has no monetary benefit. (C. Devendra and C. C. Sevilla, 2002).

Potential use of AIBPs and NCFRs on small mixed farms is enormous but contribute to less than 10% of feed in farming practice despite constituting 43% of available feed resources in Asia (C. Devendra and C. C. Sevilla, 2002).

The overall extent of use of fodder trees and shrubs in small-holder mixed farming systems is still limited (C. Devendra and C. C. Sevilla, 2002). Fodder trees and shrubs are especially significant to the nutrition of farm animals in the harsher, drier environments of the arid/semi-arid zones and the hills and upland humid parts of Asia during summer (C. Devendra and C. C. Sevilla, 2002).

Multi-nutrient block licks consisting of urea, molasses and minerals (C. De-vendra and C. C. Sevilla, 2002) can supplement basal feed residues that are deficient in energy, nitrogen (N) and macro- and microminerals. In one study in India buffalo and cattle fed considerable amounts of green forage, millet straw and concentrate, when supplemented, improved their milk yield by 30% (C. Devendra, 1997).

High-Sugar Grasses.

Recently, high-sugar perennial ryegrasses, with increased content of WSCHO, were developed from ryegrass varieties from the U.K./Belgium and Northern Italy (Anon, 2003). WSC may limit microbial protein synthesis in the rumen. The greater flow of microbial N or protein with high-sugar ryegrasses is associated with higher g microbial N per kg N input in the diet (Anon.). The understanding of the regulation of fructan, which was previously mentioned as a WSC, and the cloning of fructan metabolism genes in grasses may allow for the eventual genetic engineering in tropical grass varieties. Examples in the Philippines of shade-tolerant, tropical grasses are *Paspalum conjugatum, Setaria palmifolia* and *Centrosema accresens* in native pastures growing under coconut (O. O. Parawan and H. B. Ovalo, 1986). There are various methods including burning, overgrazing and plowing that can be used to oversow them to upgrade grasslands (F. A. Moog, 2002). These new varieties might also be selected for other qualities such as higher growth rate and better ground cover.

Varietal Selection of Feed Residues.

The improvement of the quality of cereal straws still remains unclear with breeding programmes (C. Devendra, 1997) and is proposed. It has been found that environmental factors also influence rice straw quality and that screening for digestibility may only prove useful once they are understood (D. B. Roxas et al., 1997). Increased quality would address both the digestibility of residues as well as improving crude protein content. There are technologies to down-regulate lignin content with genetic engineering and perhaps future research will reveal developmental or regulatory genes that control the protein content of residues.

Protein Enrichment.

There has been research to the manipulation of amino acids and proteins in grain with the need to better understand proteins and regulation to increase production in these feed supplements. It is proposed that research be done to manipulate the amino acids and proteins in oilseeds, forage as in straws, stems, stalks and legume haulms, green forage such as grasses, sweet potato vines, cassava leaves and fodder trees and shrubs. The potential is great for this research application as it is predicted that nutrition would be signficantly improved with N, protein and amino acid supplementation in the rumen.

Fodder Trees and Shrubs with Anti-protozoal Agents.

Animal productivity has been shown to improve with removal of protozoa or with defaunation; protozoa decreases the flow of bacterial cells to the intestines due to the predation of protozoa on rumen bacteria, digest particulate protein in the rumen with a net loss of dietary protein and are largely retained in the rumen (R. A. Leng et al., 1992). In vitro assays with various forages including *Enterolobium spp.* have anti-protozoal properties and feeding *Enterolobium cyclocarpum* has resulted in increased body weight gain and wool growth in sheep (R. A. Leng at al., 1992).

The use of effective secondary plant anti-protozoal compounds to manipulate protozoa may be engineered into various fodder trees and shrubs in the future.

Regulating Tannin Content in Fodder Trees and Shrubs.

In addition to their high levels of protein, vitamins and minerals, fodder trees and shrubs have complex chemical constituents; of the phenolics are the tannins, which at levels above 5% tannnic acid have anti-nutritional properties, and are toxic; however, at lower levels the condensed tannins may be complexed with proteins in the rumen avoiding excessive degradation (C. Devendra and C. C. Sevilla, 2002). Tannins could be down-regulated by genetic engineering.

Down-regulating Protease Activity in Forage.

When forage are harvested and/or fed upon, proteases act to breakdown proteins to endproducts including peptides and amino acids making them more soluble and leads to the inefficient breakdown by rumen microbes rather than capture of these pre-formed amino acids for greater efficiency of microbial synthesis. Down-regulating protease activity if these proteases can be identified, that are released upon cell death, and regulating them in order to avoid their expression or release. Green forages which might benefit from this manipulation are grasses, legume haulms, vines such as from sweet potato, leaves such as from cassava and fodder trees and shrubs.

Feed Resources in the Philippines.

A recent paper providing information on feed resources in the Philippines was prepared by F. A. Moog (2002) from which our discussion will follow.

The feed resource base of ruminant livestock in the Philippines consists of grasslands, the weeds, areas under plantation, primarily coconuts, residues from croplands and industrial by-products (F. A. Moog, 2002)

The Philippines has 1.5 m ha of grassland and leased through agreements; grasslands are dominated by the variety *Imperata cylindrica* constituting 50% of the total grazing area (F. A. Moog, 2002).

Improving native grasslands involves burning, overgrazing or using plowing techniques and then oversowing with pasture legumes; among the legume species that have been successfully tried and utilized are *Centrosema pubescens, Stylosan- thes guyanensis* c. v. Cook and *Leucaena leucocephala;* oversowing resulted in the improvement of the quantity and quality of herbage available resulting in better animal performance and higher productivity per unit area of land; better liveweight gain is experienced with pasture stocked with animals on native grasses and pasture legumes (F. A. Moog, 2002).

In cultivated croplands the main source of feeds are weeds that grow in culti-vated rice and corn fields including *Echinochloa colona, Rottboellia, Ischaemum rogosum* and *Dactyloctenium aegyptium;* other grasses are *Imperata cyclindrica, Paspalum conjugatum, Cyrtococcum spp.* which grow in orchards and wastelands and idlelots where broadleafs are found which also grow in fields and residues from crop production including rice straw, corn stover, mungbean hay, camote vines, cassava leaves, peanut hay, sugarcane tops and pineapple pulp; most are left to burn in the field after harvest; a 1996 crop production data report indicated 24.9 million tons of crop residues that could support 6.7 million animal units; the development and expansion of ruminant production will be dependent on improved and efficient use of crop residues (F. A. Moog, 2002). Some farmers plant forage crops either in monoculture or intercropped to provide higher quality feed and provide feed during periods of scarcity; rice and corn bran are fed when grains are milled but only rarely (F. A. Moog, 2002).

In rice-growing areas weeds constitute 50% of feeds along with rice straw and other crop residues; stall feeding and tethering of animals in uncropped and idle lands is practiced during the growing period of the rice crop; rice straw is the principal feed after rice harvest; after rice harvest fields become grazing areas in uncropped, rain-fed areas; other crop residues are corn stover and legume hays (F. A. Moog, 2002).

In sugarcane growing regions weeds are fed during the cropping season while sugarcane tops constitute 75-100% of the feeds after harvest and cane milling (F. A. Moog, 2002).

There is 3.1 m ha planted to coconut that can be utilized for livestock production; in 1983 400,000 ha of coconut land was stocked with cattle, buffalo and goats; the major components of vegetation are grasses composed of *Imperata cyclindrica, Axonopus compressus, Paspalum conjugatum* and *Cyrtococcum spp*; legumes include *Centrosema pubescens, Calopogonium mucunoides* and *Pueraria javanica*; crops are also grown under coconuts in some areas; there is considerable potential for expansion of the livestock industry with coconut including with the improvement of pasture (F. A. Moog, 2002).

Improved native pasture under coconut does not reduce coconut yield; coconut yields in grazed improved pasture were found to be higher than in ungrazed or grazed native pasture areas; guinea grass (*Panicum maximum*) and para grasses (*Brachiara mutica*) are the most common species grown under coconuts; other grasses that grow well under coconut are signal grass (*Brachiara decumbens*) and humidicola (*Brachiara humidicola*) (F. A Moog, 2002).

Arable irrigated land in larger commercial ranches can be planted and cultivated with higher yielding grasses and grass/legume mixtures; mixed grass/legume pastures are more desirable to the more costly nitrogen fertilization; sown grass and grass/legume pastures have a higher carrying capacity and provide higher liveweight gain compared to native pasture; grass/legume mixtures include para grass +centro, Napier/centro, guinea grass/Cook stylo and signal grass/Cook stylo (F. A. Moog, 2002).

Fodder trees can also be integrated with croplands and small holder farms with a need for alternative feeds in the dry season and are grown naturally in most farms; these fodder trees include *Leucaena leucocephala, Gliricidia sepium* and rain tree (*Albizia saman*) (F. A. Moog, 2002). In the system of alleycrops or intercrops tree species include *Calliandria, Leucaena, Gliricidia, Fleminga* and *Desmodium rensonii* (F. A. Moog, 2002). Leucaena has been the most popular but is not used extensively except in Batangas province where backyard cattle fattening is practiced (F. A. Moog, 2002). Sesbania is also found in rice growing areas and alongside roadsides and gardens; rain tree is fed with its leaves and pods during the dry season; gliricidia is more versatile than leucaena as a shade plant; however it is relished more by sheep and goats than by cattle; sheep attain the highest liveweight gain and efficiency per unit feed with high levels of gliricidia in their ration (F. A. Moog, 2002).

In Batangas province where fattening of one or a few head of cattle is common practice roughage in the form of fresh grass, cane tops, corn stover and rice straw is supplemented with fresh *Leucaena leucocephala*; some farmers feed concentrate fed at 0.5% of body weight consisting of copra meal, rice bran, salt, ground oyster shell and molasses fed with leucaena and cassava leaves when available (F. A. Moog, 2002).

Goats are fed supplements of Leucaena, Gliricidia and other fodder trees as well as free graze other shrubs and vegetation (F. A. Moog, 2002).

Commercial feedlots utilize sugarcane tops, banana rejects, pineapple pulp or brewers' spent grain. Green corn fodder is supplied by farmers to feedlots rather than growing corn for grain with 3-4 crops per year and avoiding crop failure with drought (F. A. Moog, 2002).

Short Summary.

Small-holder farmers constitute the majority of the farming population in developing countries in Asia. Ruminant production systems, predominantly with small-holder farmers in Asia are unlikely to change in the foreseeable future although intensification is also envisaged. Feed resources and nutrition are the principal constraints to animal production. Feed resources in Asia includes grasses, legumes and other fodders, crop residues, some of which are AIBPs, NCFRs and fodder trees and shrubs. Possible on-farm technologies applied to feeds to be discussed include chemical pre-treatment of feed residues, supplementation with feeds including multi-nutrient block licks, proposed use of varieties selected for improved straw quality with lower lignin and improved crude protein content, protein enrichment, use of forages with anti-protozoal agents and their genetic engineering in forages to improve productivity limited by protein supply, elimi-nation of anti-nutritional properties and problems of toxicity in fodder trees and shrubs due to tannin content controlled by genetic engineering, and finally high-sugar grasses high in WSC which result in increased efficiency of microbial protein synthesis in the rumen with grass varieties in coconut lands cloned with fructan (a sugar or WSC) metabolism genes. Pre-treatment and supplementation have dominated research although there is still no discernable impact on the small farm. Included are physical treatment, urea-ammonia pre-tretament of straw, use of AIBPs and NCFRs, use of secondary crops to food crops and the use of fodder trees and shrubs. Various biotechnologies can be used to address the need to improve fibre availability further including the engineering of low-lignin forages coupled with urea pre-treatment, which, it might be suggested, could possibly have a synergistic effect. There are the proposed use of anaerobic ligninolysis with ensilage or fermentation and fungal solid-state fermentation, use of enzymes (eg. lacasses) in reactors to help extract fibre in low-lignin feed residues such as straw and bagasse and the proposed use of somatic embryogenesis to increase fibre production in cereal straws and bagasse and improve standability in low-lignin, dual-purpose cereal cultivars. There is also a discussion here on the feed resources in one country, the Philippines. Poor quality native grasses together with fibrous residues form the principal feeds for ruminants on small mixed farms. Grasses are from native grasslands, rangelands, forests, fallows, wastelands, roadsides and cultivated areas after crop harvest. Pastures in Thailand are improved with grazing lands, roadsides and backyard pastures oversown with *Stylosanthes hamata, Brachiara ruziziensis, Panicum maximum* cv Hamil, *Panicum maximum* and *Pinneseteum purpureum*. In Southeast Asia there is an estimated 210 million ha for pasture improvement with perennial tree crop systems with different combinations of livestock with crops. Most attention has been given to integration with coconut. Grasses such as *Setaria splendida, Tripsacum laxum* and hybrid napier or *P. purpureum* x *P. glaucum* have been used in Malaysian rubber and oil palm plantations. Typical grass weeds associated with rice are *Echinochloa crusgalli, E. colonum, E. cruspavonis* and *Leptochloa chinensis*. Improved forages, in particular legumes, can be used with food crops in inter- and relay crops, sequence crops in rotations and improvement of fallows. Food-feed intercropping is viewed not only as a contribution to human food supplies in the short-term but also for its contribution to animal feeds and its sustainability over the long-term. In India a rice-wheat-cowpea sequence and oats, berseem and lucerne sequence and sun hemp intercropped with mung bean and pigeon pea is used. In the Philippines rice-mung bean and rice-mung bean/siratro is used. In Pakistan cow pea was intercropped with maize. In the Philippines *Sesbania* and *Desmanthus* was used with rice and cow pea and pigeon pea also used with rice. In South Asia lentils, chickpea and ground nut and lathyrus have been used with rice. In India's four northern states 30% of the cropped area irrigated farming systems have been sown with forage crops and in the Punjab in Pakistan 15% have been sown; in the summer pearl millet, maize, sorghum and cowpea and in winter berseem, lucerne rapeseed and oats are sown. In Bangladesh the legume *Lathyrus sativus* is integrated with rice on small mixed farms for dairy. In South Asia in rain-fed production systems there are large fallows that could be used to establish leguminous forages without interfering with crop production. Legumes can be used as cover-crops in perennial tree crop plantations such as rubber and oil palm plantations. Low-quality residues are a major group and are by-products from crops with crude protein of 3-4% and crude fibre content of 35-48% and include cereal straws, cereal stalks, cereal stovers, legume haulms, sugarcane tops, bagasse, cocoa pod husks, pineapple waste, coffee seed pulp and palm press fibre and form the base of feeding systems for ruminants in developing countries. The availability of rice straw is well over the requirement reflecting considerable underutilization by ruminants. Medium-quality residues including AIBPs have medium protein content including coconut cake, palm kernel cake and sweet potato vines, Good-quality residues including AIBPs are high in protein content, high in energy, minerals and vitamins and include soya bean meal, ground nut cake and cassava leaf meal. Non-traditional feeds or NCFRs include cocoa pod husks, rubber seed meal, distillers solubles, shrimp waste, leather shavings and poultry litter. Among the fodder trees and shrubs research and development has focused on *Leucaena leucocephala* and *Gliricidia sepium* in Asia and Africa and *Erythrina spp.* in Latin America. *Leucaena leucocephala* has resulted in significant increases in liveweight gain or milk yield and reduced cost of feeding due to reduced use of concentrates. If agricultural lands in Asia were planted with 230 fodder trees and shrubs per ha, this would produce 150 extra tons of fodder per hectare and support a total of 56 million animal units given as the sole diet during the dry season. The TSFS in Indonesia is planted in 0.25 ha and provides fodder throughout the year and cash crops. There is increased forage production, higher stocking rates and liveweight gains, increased farm income, reduced soil erosion and increased soil fertility and a supply of fuel wood. Another scheme is a form of upland farming called SALT utilizing alleys sown with annual and perennial crops between contoured double hedgerows of leguminous fodder trees and shrubs of *Calliandria calothysus, Leucaena leucocephala, L. diversifolia, Gliricidia sepium* and *Fleminga macrophylla*. Every third row is planted with perennial crops while others are planted with cereals, grain legumes and other annual crops. A variant SALT-2 is simple agro-livestock technology, SALT-3 is sustainable agro-forestry land technology and SALT-4 is small agro-fruit livelihood technology; SALT-2 results in a monthly net profit of USD$ 213 per ha. 23 million ha of sloping uplands are planned for development. Among treatment of cereal straws alkali urea-ammonia has been the most significant one. Variables include level of urea, compaction of straw, application by spraying or impregnation, moisture content of straw and use of open or closed containers. One study in Sri Lanka involved upgrading with impregnation versus supplementing resulted in significant higher growth rates and higher intake with upgrading. There are constraints in adopting this technology in small-scale farms including comparing the cost of inputs with the price of beef. Supplements for supplementation already mentioned, other than low quality residues, include grasses and legumes and other fodders with grasses from various areas including grazing lands and roadsides and backyard pastures, pasture improved with perennial tree crop systems, grass weeds, improved forages, in particular legumes, used with food crops in inter-, relay crops and sequence crops in rotations and improvement of fallows and as cover-crops in tree crop plantations, cakes and meals or AIBPs, fodders such as cassava leaf meal and sweet potato vines, NCFRs, in TSFS grass and herbaceous legumes, shrub legumes and fodder trees and SALT alley-cropping fodder trees and shrubs. A successful example of supplementation is with CSC in Henan province in China for beef production on a basal diet of urea-treated straw where profit increased 10-fold with the use of CSC. In rain-fed production systems of Asia adoption of food crop-feed systems in small farms is still very low; often the inclusion of forages as secondary crops to food crops have no immediate monetary benefit. Potential use of AIBPs and NCFRs on small mixed farms is enormous but contribute to less than 10% of feeds despite constituting 43% of available feed resources in Asia. The overall extent of use of fodder trees and shrubs in small-holder mixed farm systems is still limited. Multi-nutrient block licks consisting of urea, molasses and minerals can supplement basal feed residues deficient in energy, nitrogen (N) and macro- and microminerals. In a study in India with buffalo and cattle fed considerable amounts of forage and concentrate milk yield improved by 30%. Tropical grasses with increased WSC should be developed with WSC increasing the efficiency of microbial protein synthesis in the rumen and the greater the flow of microbial N or protein with high-sugar ryegrasses is associated with higher g microbial N per kg N input in the diet; understanding the regulation of fructan, a WSC, and cloning of fructan metabolism genes may allow for their eventual genetic engineering into grass varieties. Examples of shade-tolerant tropical grasses are *Puspalum conjugatum*, *Setaria palmifolia* and *Centrosema accresens* growing under coconut. Varietal selection with breeding programmes to improve straw quality is proposed although how environmental factors influences this still needs to be better understood; increased quality would address both digestibility and improving crude protein content. There are technologies to down-regulate lignin content with genetic engineering and perhaps future research will reveal regulatory genes that control protein content. There has been research towards the manipulation of amino acids and proteins in grain with the need to better understand proteins and their regulation to increase production in these feed supplements. Proposed for manipulation are straws, stems, stalks, stovers, legume haulms, green forages such as grasses, sweet potato vines, cassava leaves and fodder trees and shrubs. There is great potential in this application as nutrition would be supplemented with N, protein and amino acids in the rumen. Animal productivity has been shown to improve with removal of protozoa. In vitro assays with *Enterolobium spp.* have anti-protozoal properties and feeding *Enterolobium cyclocarpum* has resulted in increased body weight gain and wool growth in sheep. The use of secondary plant anti-protozoal compounds to manipulate protozoal growth may be engineered in fodder trees and shrubs in the future. In addition to having high levels of protein, vitamins and minerals, fodder trees and shrubs have phenolics called tannins which above levels of 5% are anti-nutritional and are toxic; however, at lower levels the condensed tannins may be complexed with protein in the rumen avoiding excessive degradation and tannins down-regulated by genetic engineering. Down-regulating protease activity would be beneficial as when forage is harvested and/or fed they breakdown proteins leading to inefficient utilization by rumen microbes and capture for greater microbial protein synthesis. Proteases that can be identified that are released upon cell death are to be regulated avoiding their expression and release. Green forages which might benefit are grasses, legume haulms, vines, leaves and fodder trees and shrubs. The feed resource base in the Philippines consists of grasslands, the weeds, areas under plantations, primarily coconuts, residues from croplands and industrial byproducts. There are 1.5 m ha of grasslands with *Imperata cylindrica* in grazing areas. Improving native grasslands involves burning, overgrazing or using plowing and then oversowing with legumes; legumes include *Censtrosema pubescens*, *Stylosanthes guyanensis* c. v. Cook and *Leucaena leucocephala*; oversowing resulted in better animal peformance or higher productivity per unit area of land. In cultivated croplands the main source of feeds are weeds that grow in rice and corn fields including *Echinochloa colona, Rottboellia, Ischaemum rogosum* and *Dactyloctenium aegyptium;* other grasses are *Imperata cyclindrica, Paspalum conjugatum* and *Cyrtococcum spp.* which grow in orchards and wastelands and idlelots where broadleafs are also found which also grow in fields and residues including rice straw, corn stover, mungbean hay, camote vines, cassava leaves, peanut hay, sugarcane tops and pineapple pulp but most are left to burn in the field; a 1996 crop production data report indicated 24.9 million tons of crop residues that could support 6.7 million animal units; the development and expansion of ruminant production in the Philippines will depend on improved and efficient use of crop residues. In rice growing areas weeds constitute 50% of feeds along with rice straw and other crop residues; stall feeding and tethering of animals in uncropped and idlelots is practiced during the growing period of the rice crop; rice straw is the principal feed after rice harvest and fields becoming grazing areas in uncropped rain-fed areas; other crop residues are corn stover and legume hays. In sugarcane growing regions weeds are fed during the cropping season while sugarcane tops are a major feed after harvest. There are 3.1 m ha planted to coconut that can be utilized for livestock production; in 1983 400,000 ha of coconut land was stocked with cattle, buffalo and goats; major grasses are *Imperata cyclindrica, Axonopus compressus, Paspalum conjugatum* and *Cyrtococcum spp.* and legumes are *Centrosema pubescens, Calopogonium mucunoides* and *Pueraria javanica.* There is considerable potential for expansion of the livestock industry with coconut including with the improvement of pasture in the Philippines. *Panicum maximum, Brachiara mutica, Brachiara decumbens* and *Brachiara humidicola* are grown under coconut. Arable irrigated land in larger commercial ranches is planted with higher yielding grasses and grass/legume mixtures. Sown grass and grass/legume pastures have a higher carrying capacity and provide higher liveweight gain compared to native pasture; grass/legume mixtures include para grass+centro, Napier/centro, guinea grass/Cook stylo and signal grass/Cook stylo. Fodder trees are also being integrated with croplands and small holder farms with a need for alternative feeds in the dry season and are grown in most farms and include *Leucaena leucocephala, Gliricidia sepium* and rain tree (*Albizia saman*). In alley-crops tree species used are *Cal-liandria, Leucaena, Gliricidia, Fleminga* and *Desmodium rensonii.* Leucaena has been the most popular but is not used extensively except in Batangas province with backyard cattle fattening. *Gliricidia* is versatile as a shade plant and is relished more by sheep and goats; sheep attain highest liveweight gain and efficiency per unit feed with high levels of *Gliricidia* in their ration. In Batangas province where fattening one or a few head of cattle is common practice, roughage consisting of fresh grass, cane tops, corn stover and rice straw is supplemented with fresh *Leucaena leucocephala* while concentrate consisting of copra meal, rice bran, salt, ground oyster shell and molasses with leucaena and cassava leaves can be fed. Goats are given supplements of *Leucaena, Gliricidia* and other fodders trees. Commercial feedlots have utilized sugarcane tops, banana rejects, pineapple pulp or brewers' spent grain. Green corn fodder is supplied by farmers to feedlots.

References.

1. Anon. 2003. Integrated research leads to award-winning grass and oat varieties. business October issue 5-6.
2. Anon. Work-package Number 3 (WP3) Improving the efficiency of rumenfunction.http://www.sweetgrassinerurope.org/Public-Pages/researchfindings. htm.
3. C. Devendra. 1997. Crop Residues for Feeding Animals in Asia: Technology Development and Adoption in Crop/Livestock Systems. In: Crop Residues in Sustainable Mixed Crop/Livestock Farming Systems. C. Renard (Ed.). Pp. 241-267. CAB International, Wallingford, Oxfordshire, U. K.
4. C. Devendra and C. C. Sevilla. 2002. Availability and Use of Feed Resources in Crop-Animal Systems in Asia. Agricultural Systems 71: 59-73.
5. R. A. Leng, S. H. Bird, A. Klieve, B. S. Choo, F. M. Ball, G. Asefa, P. Brumby, V. D. Mudgal, U. B. Chaudry, S. U. Haryono and N. Hendratno. 1992. The Potential for Tree Forage Supplements to Manipulate Rumen Protozoa to Enhance Protein to Energy Ratios in Ruminants Fed On Poor Quality Forages. In: Legume, Trees and Other Fodder Trees as Protein Sources for Livestock. Vol. 102. A. Speedy and P. L. Pugliese (Eds.). Pp. 177-191. FAO, Rome, Italy.
6. F. A. Moog. 2002. Country Pasture/Forage Resources Profiles. Livestock and Development Council.
7. D. B. Roxas, M. Wanapat and Md. Winugroho. 1997. Dynamics of Feed Resources in Mixed Farming Systems in Southeast Asia. In: Crop Residues in Sustainable Mixed Crop/Livestock Farming Systems. C. Renard (Ed.). Pp. 101-112. CAB International, Wallingford, Oxfordshire, U. K.

Chapter 9

BIOTECHNOLOGY AND FODDER
TREES AND SHRUBS

The Problem and the Potential.

Feeds for animal production have been proposed with technologies to make them readily available to the small farmer at the small-farm level with on-farm technology derived from technology from the research station or at the experimental development stage.

Various approaches have been proposed including management of feeding practices such as pre-treatment of feed residues or fibre in forage, the utilization by feeding of agro-industrial byproducts (AIBPs) such as cereal straws and non-conventional feed resources (eg. chicken litter as a source of crude protein) (NCFRs), use of secondary crops including food crops such as legumes to supplement feed crops and use of fodder trees and shrubs planted with other crops including cash crops.

The breeding or the genetic engineering in plants has also been proposed in regards to agronomic traits for food or feed production, improving the compositional qualities of plants as food or feed and improving the qualities in plants that affect utilization of feeds in animals.

We will discuss various applications that biotechnology can provide to improve the utilization of fodder tree and shrub species which have been proposed and is expected to be a more widespread feed resource for livestock in future. The qualities manipulated in fodder trees and shrubs will be low-lignin content, high-sugar content, down-regulated protease activity, the introduction of anti-protozoal activity (protozoa areside as normal fauna in the rumen of livestock), to increase ruminal microbial protein synthesis, the protection of protein to escape rumen digestion which is practical when the protein in the feed source is of especially high protein value and is to escape for optimum supplementation of feed to the animal with more intensive production, the further enrichment of leaf protein with the introduction of protein from trees and shrubs (e.g. Rubisco, a ubiquitous protein in plants) and other sources attempted as 'surrogates' (e. g. albumins) and introduction of drought tolerance, herbicide tolerance, pest resistance and disease resistance.

Low-lignin Forage.

Lignin down-regulation has been proposed and developed experimentally in poplar plants using sense repression, using DNA in the same orientation, and anti-sense repression, using DNA in the opposite orientation, of major biosynthetic enzymes such as caffeic acid O-methyltransferase (COMT), caffeoyl coenzyme A O-methyltansferase (CCoAOMT) and other enzymes such as 4-coumarate:CoA ligase (4CL), cinnamoyl CoA reductase (CCR), cinnamyl alcohol dehyrdogenase (CAD) and ferulate-5-hydroxylase (F5H). In poplar trees according to its biosynthetic schemata, given assumptions of which enzymes are active or blocked, key enzymes, and which alternative enzymes could be present in larger quantity or magnitude of activity, down-regulating CCoAOMT has resulted in reduced lignin content (see: Chapter 4). Forage such as low-lignin alfalfa has already been developed with modified lignin and cellulose content; however, the problem of stem strength and standability still needs to be resolved (A. V. Deynze et al., 2004) but this has not been a problem with poplar trees unlike alfalfa. Fodder trees and shrubs may prove to have good standability with increased digestibility in their leaves as fodder. Other forage grasses such as tall fescue and bermuda grass and the legume white clover have been reported with research to manipulate lignin to improve digestibility (Z. Wang, 2003).

High-Sugar Forage.

Water-soluble carbohydradates (WSC), including fructans, occur in various plant species and has improved ruminal microbial digestion in ruminants. It is still not known what mechanism(s) are involved. One theory proposed is that of microbial protein synthesis resulting in greater microbial nitrogen (N) flow from the rumen. One study with high-sugar and control grasses as silage with red clover silage resulted in lower rumen ammonia-N concentrations and higher microbial N flow to the small intestine (Anon; undated). It has been proposed that high-sugar grasses be developed in the tropics where grazing lands under tree plantations could be used with intensive dairy goat production. Fodder trees and shrubs as browse plants as sources of protein for ruminant livestock would benefit from the addition of sugars further boosting ruminal protein production.

The introduction of enzymes for fructan biosynthesis in other plant hosts has been successful. There are three fructan types. The inulins are synthesized with sucrose to which is added a fructose unit beta(2->1) to it to form 1-kestose with sucrose:sucrose 1-fructosyltransferase (1-SST) and elongated beta (2->1) with fructan:fructan 1-fructosyltransferase (1-FFT) (T. Ritsema and S. C. M. Smeekens, 2003). Examples of gene transfer to plant crops is 1-SST and 1-FFT from globe artichoke to potato (T. Ritsema and S. C. M. Smeekens, 2003). Levan, the second type of fructan, is 1-kestose with beta(2->1) branches and elongated with beta(2->6) to fructose units with sucrose:fructan 6-fructosyltransferase (6-SFT); 6-SFT can be used to form 1-kestose from sucrose (T. Ritsema and S. C. M. Smeekens, 2003). When 6-SFT is transferred to chicory where inulins occur a mixture of levan and inulin fructans result (T. Ritsema and S. C. M. Smeekens, 2003). The third type graminans are a combination of fructose linked beta(2->1) and beta(2->6) to 1-kestose which occurs in grasses (T. Ritsema and S. C. M. Smeekens, 2003). The last type of fructan are the inulin neoseries which is based on neokestose which is sucrose attached to fructose via beta(2->6) linkage from which, with fructan:fructan 6G-fructosyltransferase (6G-FFT), beta(2->1) and beta(2->6) chains can be elongated with fructose at the two fructose ends (T. Ritsema and S. C. M. Smeekens, 2003). 6G-FFT from onion was transferred to chicory which resulted in fructan of the inulin neoseries type occurring in chicory (T. Ritsema and S. C. M. Smeekens, 2003).

It is speculative at this time, but possible, that donors can be used to transfer fructosyltransferase genes to other species such as tree crop plantation grasses, sugarcane and fodder trees and shrubs. There would be significant benefits as a result as has been reported with the breeding of high-sugar ryegrasses.

Down-regulated Protease in Forage.

There are no known reports of manipulation of protease activity in grasses or legumes or other species such as fodder trees and shrubs. Normally proteases act to breakdown proteins with cell death or after injury or damage. It will be necessary to isolate and further study these enzymes first. The theory of how protease activity affects digestion of plant proteins in the rumen has been put forward whereby decreasing plant proteases still is presented with the fact of the rumen and its well adapted microbial proteolytic action on nitrogen (N)-rich feeds. It is of interest whether down-regulating proteases further to pre-wilting with heat damage to proteins make protein less digestible and increasing supply of peptides and amino acids to microbial protein synthesis resulting in increased protein flow from the rumen.

Anti-protozoal Compounds in Forage.

There are anti-protozoal forages with potential anti-protozoal-active compounds that might be developed for forages where protein supplementation might be required such as with sugarcane, mature grasses and low-quality silages and/or for more intensive production.

Protozoa naturally occur in the rumen; because of predation of bacteria in the rumen, their retention in the rumen and digestion of particulate protein there is less flow of protein to the intestines (R. A. Leng et al., 2003). An example of forage with anti-protozoal activity is *Enterolobium cyclocarpum* (E. C.); when E. C. is used as a supplement there is a body weight gain of 24% and wool growth gain of 27% in sheep (R. A. Leng et al., 1992).

Protecting Protein in Forage.

Condensed tannins are phenolic compounds that can form protein-tannin complexes when present, for eg. condensed tannins at 2-3%; the complexes form depending on pH, ionic strength and molecular size (complexation usually at M. W. <5000); condensed tannins as opposed to hydrolysable tannins are more effective in precipitating proteins (R. Kumar, 1992). Protein-tannin complexes can form with protein in the rumen as well as with endogenous protein such as pepsin;

the complex appears to dissociate post-ruminally at a low pH where protein becomes available (R. Kumar, 1992).

When protein sources in the diet of livestock are of high-quality and because of inefficient breakdown in the rumen, when optimized, it is best to escape rumen degradation; this is where the use of tannins at low levels with feeds might be useful such as corn and fodder trees and shrubs which are potential supplements.

Enrichment of Protein.

Cereals have been investigated as protein sources that might be improved in amino acid and protein content. Barley is a good example that has been studied and certain classes of proteins identified with the approach that certain proteins high in specific essential amino acids are selected over those low in these essential amino acids. However their regulation has to be further studied. No recent reports of this type of research have been made but it is expected that there will be progress and that high-yielding, high-protein varieties in cereals such as corn, barley and wheat will result. For fodder trees and shrubs further boosting production in leaf proteins should be possible. Major classes of protein that can be used both from trees and shrubs and other sources have to be identified with their genes isolated and their function identified together with their regulatory sequences in order to genetically engineer them.

Other Traits.

Researchers have already previously investigated traits for drought tolerance, herbicide tolerance, pest and disease resistance with the use of genetic engineering.

Drought tolerance and resistance to freezing imply a relationship resulting in dehydration; researchers have isolated genes being turned on and use them to identify enzymatic players in stress response; an example are cold-regulated or COR genes turned on by drought conditions via overexpressing transcription factors or regulators called CBF/DRB regulon proteins; genetic engineering here involves manipulation or regulation of whole complex pathways via a master switch rather than just via an individual "end player" protein; in manipulating stress tolerance there may be plieotropic effects, ie. effects on production of plants (D. Amber, 2000). A report of drought resistance with other stacked traits (N capture from ammonium-based fertilizers, resistance to a weed killer and changes in amino acid composition) in corn has been proposed in research at Southern Illinois University USA. It is debatable as to the need for drought tolerance in fodder trees and shrubs in sub-tropical and tropical climes but sub-Saharan Africa could be more likely.

Tolerance to the herbicides such as Round-Up ® (glyphosate) and Liberty ® in the USA have been developed. They can be introduced in tropical areas where fodder trees and shrubs might benefit minimizing the need for labour intensive maintenance of crops.

A well-known resistance mechanism with the European corn borer in corn is with the genetically engineered insecticidal proteins that affect the borer's gut from the soil bacterium *Bacillus thuringiensis* or Bt; there could be a future problem of insect resistance although it has been proposed that use of refuges with non-genetically altered corn be grown together with the Bt-added corn so that insects can dilute any species that develop resistance to it; also several insecticidal proteins can be used making available resistance less likely (D. Tenenbaum, 1998).

Sense-repression where the transgene in the same direction silences transcripts of the gene complementary to it, ie. the arrangement of the double-stranded genes would be in opposite orientations, has been used. Squash, mango and sweet potato are well-known examples with viral resistance using coat protein transgenes from viruses. Another term used for this technique is post-transcriptional gene silencing where there is a relatively low level of accumulation of coat protein mRNA and no or little protein product (Anon., 2000).

Short Summary

Feed technologies for the small farm or on-farm technologies have been proposed. Feed management practices are pre-treatment of feed residues or fibre, the utilization of AIBPs such as cereal straws and NCFRs (eg. chicken litter), use of secondary crops including food crops to supplement feed crops and use of fodder trees and shrubs planted with other crops including cash crops. The breeding or genetic engineering in plants has been proposed in regards to agronomic traits, compositional qualities and those affecting utilization of feeds in animals. Discussed are various applications of biotechnology to fodder trees and shrubs expected to be a more widespread feed resource in future. Qualities include low-lignin content, high-sugar content, down-regulated protease activity, anti-protozoal activity to increase ruminal microbial protein synthesis, protection of protein to escape rumen digestion for optimum supplementation with more intensive production, the further enrichment of leaf protein with introduction of proteins introduced from trees and shrubs and other sources and introduction of drought tolerance, tolerance to herbicides, pest resistance and disease resistance. Lignin down-regulation has been proposed and developed in poplar plants using sense repression and anti-sense repression of major biosynthetic enzymes such as COMT, CCoAOMT and other enzymes such as 4CL, CCR, CAD and F5H. With certain assumptions in lignin biosynthetic pathways the down-regulation of CCoAOMT has resulted in reduced lignin content. Low-lignin alfalfa has already been developed with modified lignin and cellulose content, however, with stem strength and standability still needing to be resolved unlike poplar trees. Fodder trees and shrubs may prove to have good standability with increased digestibility in their leaves. Reports have been made on research to study the manipulation of lignin in forage grasses such as tall fescue and bermuda grass and the legume white clover. WSC, including fructans, occur in various plant species and improves ruminal microbial digestion in ruminants. The mechanism(s) are not known although one theory proposed is that of greater microbial N flow from the rumen. One study with high-sugar and control grasses as silage and red clover silage resulted in lower rumen ammonia-N concentration and higher microbial N flow to the small intestines.

It has been proposed that high-sugar grasses be developed in the tropics where grazing lands under tree plantations be used with intensive dairy goat production.

Addition of sugars would further boost ruminal protein production from fodder trees and shrubs as sources of protein for ruminant livestock. The introduction of fructan biosynthesis in other plant hosts has been successful. There are four types of fructans: the inulins with 1-SST which synthesize 1-kestose and elongated beta(2->1) with 1-FFT; 1-SST and 1-FFT has been transferred from globe artichoke to potato; the second type of fructan is levan from 1-kestose with beta(2->1) branches and elongated beta(2->6) with 6-SFT; introduction of 6-SFT in chicory has resulted in a mixture of both inulins and levans; the third type of fructan are the graminans which are a combination of fructose linked beta(2->1) and beta- (2->6) to 1-kestose in grasses; the last type are the inulin neoseries based on neo- kestose formed with 6G-FFT with beta(2->1) and beta(2->6) elongations; 6G-FFT from onion transferred to chicory result in the inulin neoseries in chicory. It is possible that the transfer of fructosyltransferases in tree crop plantation grasses, sugarcane and fodder trees and shrubs would result in significant benefits as with breeding of high-sugar ryegrasses. There are no known reports of the manipulation of protease activity in grasses or legumes or other species at this time. It will be necessary to isolate and study these enzymes. The theory of how plant protease activity affects digestion of plant proteins in the rumen still is presented with the fact of the well-adapted rumen microbial protease activity on N-rich diets. There is a question as to whether down-regulating protease activity further to pre-wilting with heat damage makes protein less degradable and increases supply of peptides and amino acids to microbial synthesis and ruminal protein flows. There are anti-protozoal forages with potential anti-protozoal-active compounds that might be introduced in forages where protein supplementation might be required such as with sugarcane, mature grasses and low-quality silages and/or for more intensive production. Protozoa naturally occur in the rumen. Because of predation of bacteria in the rumen, their retention in the rumen and digestion of particulate protein, less flow of protein results. An example of forage with anti-protozoal activity is *E. C.* which when used results in a body weight gain of 24% and wool growth gain of 27% in sheep. Condensed tannins are phenolic compounds that can form protein-tannin complexes when present, for example, at 2-3%. The complexes form depending on pH, ionic strength and molecular size (complexation usually at MW <5000). Condensed tannins are more effective than hydrolysable tannins in precipitating proteins; both protein complexes in the rumen and with endogenous protein and dissociate post-ruminally at low pH. Where protein is of high-quality and where there is inefficient breakdown in the rumen, it is best to escape rumen degradation and tannins at low levels such as with corn and fodder trees and shrubs are useful. Amino acid and protein content might be increased as has been already been initially investigated in barley where proteins high in specific essential amino acids are to be selected over those low in the same amino acids. No recent reports have been made on further developments but it is expected that there will be progress and that high-yielding, high-protein varieties in cereals such as corn, barley and wheat will result. The boosting of leaf proteins in shrubs and trees should be possible with major classes of proteins from shrubs and trees and other sources with their genes isolated, their function identified together with their regulatory sequences to geneticall engineer them. Other traits that have been investigated are drought tolerance related to cold stress examples of genes of which are cold-regulated or COR genes turned on by drought via overexpressing transcription factors or regulators called CBF/DRB proteins; manipulation here involves whole complex pathways via a master switch; manipulation of stress tolerance may have plieotropic effects. A report of drought resistance with other stacked traits with N capture, herbicide resistance and amino acid composition in corn have been reported. There may not necessarily be a need for drought tolerance in fodder trees and shrubs in subtropical and tropical climes but areas such as in the sub-Sahara could be likely. Herbicide tolerance to Round-Up ® and Liberty ® have been developed and introduced in tropical areas with benefits of minimizing labor-intensive maintenance of crops. Pesticide resistance in crops, e.g. Bt, is well-known; approaches such as use of refuges to control insect resistance to Bt crops and use of several insecticidal proteins to minimize the likelihood of resistance are proposed. Viral disease resistance using sense-repression are with the introduction of viral coat protein transgenes with squash, mango and sweet potato as well-known examples; another term used for this techniques is post-transcriptional gene silencing.

References.

Anon. 1998. Field of Genes. The Why Files. Internet document: http:// whyfiles.org /062ag_gene-eng/1.html.

Anon. Work-package Number 3 (WP3). Improving the efficiency of rumen function. Internet document: http://www.sweetgrassineruope.org/ Public_Pages/researchfindings.htm.

D. Amber 2000. Genetic reponses to drought. The Scientist 14: 18.

A. V. Deynze, K. J. Bradford and A. V. Eenennaam. 2004. Crop Biotechonology: Feeds for livestock. Publication 8145. Agriculture And Natural Resources. University of California, Davis, USA.

R. Kumar. 1992. Anti-nutritional factors, the potential risks of toxicity and methods to alleviate them. In: Legume Trees and Other Fodder Trees As Protein Sources for Livestock. A. Speedy and P. L. Pugliese (Eds.). FAO: Rome, Italy.

R. A. Leng, S. H. Bird, A. Klieve, B. S. Choo, F. M. Ball, G. Asefa, P. Brumby, V. D. Mudgal, U. B. Chaudhry, S. U. Haryono and N. Hendratno. 1992. The potential for tree forage supplements to manipulate rumen protozoa to enhance protein to energy ratios in ruminants fed on poor quality forages. In: Legume Trees and Other Fodder Trees as Protein Sources for Livestock. A. Speedy and P. L. Pugliese (Eds.). FAO: Rome, Italy.

T. Ritsema and S. C. M. Smeekens. 2003. Engineering fructan metabolism in plants. Journal of Plant Physiology 160: 811-820.

Z.-Y. Wang. 2003. Biotechnology has potential for forage improvement. AgNews&Views. The Noble Foundation, USA.

S. Young. Sweet grass, contented cows. Internet document.

R. Zhong, W. H. Morrison III, D. S. Himmelsbach, F. L. Poole II And Z.-H. Ye 2000. Essential role of caffeoyl coenzyme A O-methyl -transferase in lignin biosynthesis in woody poplar plants. Plant Physiology 124: 563-578.

Chapter 10

THE MANIPULATION OF FRUCTAN METABOLISM

The Problem and the Potential.

Fructans have been recognized as a class of carbohydrates for over 150 years and their chemistry, biochemistry and physiology have been studied intermittently over that period (C. J. Pollock and A. J. Cairns, 1991).

Fructans have recently been recognized as a functional food (foods that have a particular nutritional value and can be used as nutritional supplements); it reduces insulin, cholesterol and triacylglycerol and phospholipid levels in blood and can be used as an artificial sweetener and fat replacer; it can also be utilized as a bioplastic and feedstock for ethanol and glycerol production (Anon., 1996, T. Ritsema and S. C. M. Smeekens, 2003).

It may be that fructans serve as storage carbohydrates. In perennial species accumulation continues throughout Fall and Winter and concentrations decline prior to rapid spring growth (C. J. Pollock and A. J. Cairns, 1991). It has been shown that treatments that lead to a decrease in the demand for fixed carbon: temperature stress, withholding minerals, nutrient starvation [eg. nitrogen (N) and phosphorous (P)], water stress, applying growth retardants, defoliation and alternatively increasing carbon supply by increasing leaf area, photoperiod, irradiance and carbon dioxide concentration results in fructan accumulation (Anon. (undated), C. J. Pollock and A. J. Cairns, 1991).

Fructans, part of the class of water-soluble carbohydrates (WSCHOs) found in grasses and legumes, have been boosted in high-sugar ryegrasses bred using varieties in the U. K./Belgium and Italy.

Studies with high-sugar ryegrass varieties with ruminant livestock result in increased efficiencies in digestibility of forage and microbial protein synthesis and increased milk production. However, it is still unknown as to what exact mechanism(s) are involved in improving ruminant digestion.

Beyond the obvious application to feeds, the manipulation of fructan metabolism will also be applied to fruits and vegetables as functional foods competing with alternative modes of intake such as supplements but which have the advantage of reduced cost and the advantage of ease of convenience or use.

Fructans have been suggested for their added sweetness and supposed natural, delicate flavour in dishes such as those for spinach, artichoke, sweet red chili peppers, mushrooms, zucchini, avocadoes, onions, grapefruit and prepared grapefruit juices. These examples of 'fruit and veg' functional foods are believed to be worth in the billions of USD$. The popular eating banana has also been proposed for this market.

The Structure of Fructans.

There are three types of fructans with varying degrees of polymerization (DP). The sucrose molecule (with a hexose glucose unit attached via C1 to C2 of the other fructose unit) (glucose is a 6C ringed sugar (5Cs and 1O) with a C6 tail facing above the plane while fructose is a 5 ring (5C) sugar with 2 tails facing above and below the plane at C6 and C1) is the starting unit which can have beta(2->1) fructose-fructose (in the starting sucrose unit) bonds forming linear inulin; this type of fructan is found in the family Astarales (eg. chicory, Jerusalem artichoke) (T. Ritsema and S. C. M. Smeekens, 2003). Another type of fructan is levan with mainly beta(2->6) fructose-fructose bonds (C2 of fructose to C6 of fructose in the starting sucrose unit) found in grasses in the leaf base (T. Ritsema and S. C. M. Smeekens, 2003). There is also a mixed variant with beta(2->1) branches with linear beta(2->6) chains found in Poaceae (eg. wheat, barley and grasses) (T. Ritsema and S. C. M. Smeekens, 2003). A third type of fructan are the graminans with both beta(2->1) and beta(2->6) linkages found in grasses (Anon., 1996). The last type are the inulin neo-series. The starter unit is neokestose consisting of sucrose with C2 of fructose linked to C6 of the glucose unit; this is found in Liliaceae (T. Ritsema and S. C. M. Smeekens, 2003).

The Biochemistry of Fructan Metabolism.

The basic model presented in the literature for fructan biosynthesis is that of 2 enzymes: 1-SST or sucrose:sucrose 1-fructosyltransferase; as the name implies, it transfers the C1 fructosyl unit of one sucrose to C2 of fructose of the other sucrose molecule to form the trisaccharide 1-kestose; the second enzyme, 1-FFT or fructan:fructan 1-fructosyltransferase then elongates or polymerizes 1-kestose or fructans of higher DP transferring fructosyl units at C1 to C2 of the growing chain; sucrose can also be used as an acceptor by 1-FFT; higher concentrations of 1-SST leads to synthesis of longer fructans implying that it can synthesize other fructans than 1-kestose only (T. Ritsema and S. C. M. Smeekens, 2003). In bacteria inulin of much higher molecular weight is also synthesized with only one fructosyltransferase enzyme (T. Ritsema and S. C. M. Smeekens, 2003). In case of beta(2->6) linkages in levans and graminans sucrose:fructan 6-fructosyltransferase (6-SFT) is the fructosyltransferase which attaches the C2 of a fructose unit of sucrose to the C6 tail of fructose of 1-kestose to form bifurcose (T. Ritsema and S. C. M. Smeekens, 2003). 6-SFT can also form the trisaccharide 6-kestose from one sucrose unit and fructose of another sucrose unit in a beta(2->6) linkage (T. Ritsema and S. C. M. Smeekens, 2003). 6-SFT is also able to act as a 1-SST and form 1-kestose and has invertase activity (T. Ritsema and S. C. M. Smeekens, 2003). The biosynthesis of the inulin neoseries starting with neo-kestose, a trisaccharide made by 6G fructosyltransferase (6G-FFT) attaches a fructose unit from 1-kestose and low DP inulins at C2 to the C6 tail of glucose; chain elongation can be either way in beta (2->1) from neo-kestose C2 to C1 at one end and C2 to C1 at the other end with fructose (T. Ritsema and S. C. M. Smeekens, 2003). In Oat, a very complex mixture of fructans are formed with neokestose elongated at both ends with beta (2->1) and beta(2->6) linkages (T. Ritsema and S. C. M. Smeekens, 2003). 6G-FFT is also able to act as a genuine 1-FFT and form higher DP fructans (T. Ritsema and S. C. M. Smeekens, 2003).

The breakdown of fructans (ie. degradation) is accomplished by fructan exo-hydrolases (FEHs) which removes fructosyl units successively (T. Ritsema and S. C. M. Smeekens, 2003).

Invertases (enzymes that degrade sucrose to fructose and glucose) are able to degrade inulins of low DP and synthesize low DP inulins in the presence of high concentrations of sucrose (T. Ritsema and S. C. M. Smeekens, 2003). Invertases and 1-SST overlap activities in that they are able to degrade sucrose and synthesize low DP inulins (T. Ritsema and S. C. M. Smeekens, 2003).

1-SST is induced with higher sucrose concentrations as with cold, drought stress in leaves and roots and continuous light unlike 1-FFT. 1-SST in barley is induced by sucrose while 6-SFT is induced by sucrose, cold, nitrogen deficiency and light (T. Ritsema and S. C. M. Smeekens, 2003). Cold induces 1-FEH I and 1-FEH II in chicory while defoliation induces only 1-FEH II. Gibberelins might induce FEH activity and breakdown of fructans (T. Ritsema and S. C. M. Smeekens, 2003).

Genetic Engineering of Fructan Metabolism.

It is already possible to transfer fructan genes into other plant hosts (T. Ritsema and S. C. M. Smeekens, 2003).

Engineering of fructan enzymes to further boost synthesis with mixed fructans or when fructan is not present has been successful (T. Ritsema and S. C. M. Smeekens, 2003).

Characterization of promoter regions for fructan metabolism genes is with PCR-based genome walking techniques and the use of the reporter gene beta glucuronidase (GUS) with histochemical characterization.

The 1-SST and 6-SFT upstream regions from barley was characterized with use PCR to characterize and map by reiterative processes overlapping sequences and with 6-SFT use of transient expression systems that drive GUS reporter genes (genes are propelled via gold particles into cells) and histochemically characterizing them in particular plant tissue (Anon., 2004).

1-SST and 1-FFT from glove artichoke (of Astarales) has been transferred to potato and tubers and fructosyltransferases have been transferred to sugar beet resulting in fructan production (T. Ritsema and S. C. M. Smeekens, 2003).

Another approach to boosting fructan metabolism might be to downregulate FEH if it is present in sufficient amounts although it is expected that most tropical grasses and sugarcane will have little fructan content as C4 species have no fructan or have only low concentrations of it.

Short Summary

Fructans as a class of carbohydrates have been recognized and studied for over 150 years. They are a functional food reducing insulin, cholesterol and triacylglycerol and phospholipids levels in blood, used as an artificial sweetener, fat replacer, bioplastic and for ethanol and glycerol production. It may be that fructans serve as storage carbohydrates in plants. In perennial species there is accumulation in the Fall and Winter, declining prior to Spring growth. Treatments that lead to a decrease in fixed carbon and increases in carbon supply result in fructan accumulation. Some examples are temperature stress, water stress, increased photoperiod and carbon dioxide concentrations. Apart from the apparent applications to feeds, fructans has also been proposed for fruits and vegetables compared to health food supplements and suggested for specific fruits and vegetables for their supposed sweet and delicate flavours. The value of this 'fruit and veg' market is estimated in the billions of USD$ and so it also is proposed for the popular eating banana. Fructans, part of the class of water-soluble carbohydrates have been boosted in high-sugar ryegrasses. Studies with high-sugar ryegrass varieties with ruminant livestock result in increased efficiencies in digestibility, microbial protein synthesis and milk production although it is not known as to what exact mechanisms are involved. Fructan structure vary in degree of polymerization (DP). There are three types: inulin which starts with a sucrose molecule (composed of glucose and fructose) with beta(2->1) fructose-fructose linear bonds found in the family Asterales; the second is levan starting with a sucrose unit attached beta(2->1) to a fructose unit forming 1-kestose with beta (2->6) fructose-fructose bonds found in grasses. A mixed variant with beta(2->1) branches with linear beta (2->6) chains is found in wheat, barley and grasses. A third type are the graminans with both beta(2->1) and beta(2->6) linkages in grasses. The last type are the inulin neoseries. The starter unit is neokestose consisting of fructose linked to glucose of the sucrose unit with beta(2->6) and beta(2->1) branching on either side with fructose units. The biosynthesis of fructans involve 1-SST forming 1-kestose from sucrose and fructose from sucrose with elongation by 1-FFT further forming beta(2->1) bonds; other fructans can be used as donors of fructose. 1-FFT can also use sucrose as an acceptor molecule. The presence of higher concentrations of 1-SST leads to longer fructans implying that it can synthesize them other than 1-kestose. In levans and graminans 6-SFT attaches a fructose unit of sucrose to 1-kestose to form bifurcose with beta(2->6) elongations. 6G-FFT attaches a fructose unit from 1-kestose and low DP inulins to the glucose unit of sucrose; chain elongation can be beta(2->1) on either ends with fructose or beta(2->1) and beta(2->6) with fructose on either ends in Oat. The breakdown of fructans is via fructan exohydrolases (FEHs) which removes fructosyl units successively. 1-SST is induced with higher sucrose concentrations as with cold, drought stress in leaves and roots and continuous light unlike 1-FFT. 1-SST in barley is induced by sucrose while 6-SFT is induced by sucrose, cold, nitrogen deficiency, and light. Cold induces 1-FEH I and 1-FEH II in chicory while defoliation induces only 1-FEH

II. Gibberelins might induce FEH activity and breakdown of fructans. It is already possible to transfer fructan genes into other plant hosts. Engineering fructan enzymes to further boost synthesis of mixed fructans or where fructan is not present has been successful. Characterization of promoter regions would involve use PCR-based genome walking techniques and use of the reporter gene beta-glucuronidase (GUS) and histochemical characterization. The 1-SST and 6-SFT upstream regions were characterized and mapped by reiterative processes of overlapping sequences and use of transient expression systems that drive GUS reporter genes and histochemically characterizing them in particular plant tissue. 1-SST and 1-FFT from globe artichoke has been transferred to potato and tubers and fructosyltransferases have been transferred to sugar beet resulting in fructan production.

References.

1. Anon. (undated). Internet document: http://www.kuleuven.ac.be/bio/dev/research.htm.
2. Anon. 1996. Internet document: http://www.geocities.com/CapeCanaveral/4409/fructan.htm.
3. Anon. 2004. http://pages.unibas.ch/bothebel/people/vinay/vinay.htm.
4. C. J. Pollock and A. J. Cairns. 1991. Fructan Metabolism in Grasses and Cereals. Annual Review of Plant Physiol. and Plant Mol. Biol. 42: 77-101.
5. T. Ritsema and S. C. M. Smeekens. 2003. Engineering Fructan Metabolism in Plants. J. Plant Physiol. 160: 811-820.

Chapter 11

ENZYMES AND FERMENTATION
IN THE PROCESSING OF BAGASSE AS FEED

The Problem and the Potential.

Bagasse is the fibrous residue milled by-product of the sugarcane plant. It comprises about 28% of the cane plant (E. T. Baconawa, 1986) and can be used for animal feeding. It was earlier pointed out that treatment of bagasse with chemicals is expensive but treatment with enzymes or fermentation has great promise (E. T. Baconawa, 1986). It was not further elaborated upon on what course of action this would take, for example, with yeast or other microbial source.

In this paper we will discuss first the problem with the theory of phenolic compound-mediated lignin carbohydrate complexes (PC-LCCs) in the cell wall in feed particles in the discussion of Wang and McAllister (2002) that limit microbial colonization on the surface of feed particles and limit the extent of digestion of feed in the rumen. It has been suggested that the use of extracellular fibrolytic enzymes (EFEs) do not add novel enzymatic activities and is limited in their extent of digestion in the rumen (Y. Wang and T. A. McAllister, 2002).

We will discuss two industrial applications that serve as methods to pre-treatment that address the problem with the use of the recent discovery with lignases, namely, lacasse with use of mediator compounds for delignification in pilot pulp paper which may be extended to forage and the use of the solid state fermentation (SSF) process that utilizes white rot fungi (WRF) to degrade the lignocellulose along with other practical methods of delignification.

We will then discuss the use of feeding EFE preparations after the proposed feed pre-treatments. We will discuss various factors that may affect response to EFEs such as the sources of EFEs, method of application, use of different diets and level of animal production (K. A. Beauchemin et al., 2002, Y. Wang and T. A. McAllister, 2002).

Limitations to Feed Digestion by Ruminal Microbes.

The residue of the sugarcane plant, bagasse, is fibrous and highly lignified. It is of low digestibility (35%) (R. A. Leng and T. R. Preston, 1985). PC-LCCs at the surface of feed particles act both as a physical and biochemical layer that limits microbial colonization and protects the underlying layers from further attack (Y. Wang and T. A. McAllister, 2002). In support of this is the evidence that free phenolic acids and soluble phenolic carbohydrate complexes inhibit microbial activity and concentrations of these materials on the surface prevent microbial attachment (Y. Wang and T. A. McAllister, 2002). The efficiency of cell wall degradation will depend on the rate of microbial colonization and microbial removal of PC-LCCs from the particle surface to expose the contained polysaccharides (Y. Wang and T. A. McAllister, 2002).

It was proposed that EFEs enhance microbial colonization by producing reducing sugars that attract secondary colonization and remove barriers to microbial attachment by cleaving linkages between phenolic compounds and polysaccharides (Y. Wang and T. A. McAllister, 2002). Studies have shown that EFEs are limited in their ability to cleave phenolic compounds from feed particles and although reducing sugars from hydrolysis increase microbial adhesion, extensive hydrolysis reduces colonization compared to minimally hydrolyzed straw (Y. Wang and T. A. McAllister, 2002). This brings us back to the point that EFEs may be hydrolyzing the polysaccharides of the cell wall and leaving the PC-LCCs on the surface to block colonization (Y. Wang and T. A. McAllister, 2002). With in situ dry matter (DM) disappearance with the application of EFEs, the percentage of phenolic compounds in the residue increases (Y. Wang and T. A. McAllister, 2002). These are consistent with the general observation that EFEs usually only increase the rate and not the extent of digestion (Y. Wang and T. A. McAllister, 2002).

It has been observed that EFEs are limited in cleaving PC-LCC esterified linkages (Y. Wang and T. A. McAllister, 2002); research suggests that esterified linkages are cleaved by acetylesterase and ferulic acid esterase (Y. Wang and T. A. McAllister, 2002). Bonds between lignin and hemicellulose are with the residues of the hydroxycinnamic acid ferulic acid and the uronic acid 4-O-methyl-beta-D-glucuronic acid in graminaceous plants and ether bonds between lignin and the carbohydrate moiety (K. E. Hammel, 1997, Y. Wang and T. A. McAllister, 2002) and lignin composed largely of arylglycerol-beta-arylether structures constituting 50% of the monomers (K. E. Hammel, 1997). Novel esterase and etherase activities need to be identified and produced along with EFEs currently marketed (Y. Wang and T. A. McAllister, 2002). Other methods used with EFEs mentioned later together with the industrial applications discussed are alkaline pre-treatment which cleaves esterified bonds in the PC-LCC matrix improving access by microbial enzymes and increasing adhesion and colonization and steam pre-treatment which improves the hydrolysis efficiency of EFEs (Y. Wang and T. A. McAllister, 2002) with steam pre-treatment separating out the hemicellulose, cellulose and lignin components of fibre (Orskov, 2000). The use of alkali ammoniation with urea-ammonia with cereal straws in Asian countries may be optimized for maximal efficiency with use of EFEs (Y. Wang and T. A. McAllister, 2002). Both steam treatment and urea-ammonia are considered practical approaches for developing country settings.

The Upgrading of Bagasse with Industrial-scale Enzymatic and Fermentative Processing.

The processing of bagasse at industrial-scale for a developing country has been proposed in the past for the Philippines by a buyer representing a Japanese company for a fermentative process for bagasse (E. T. Baconawa, 1985). Presumably, bagasse after processing can be resold to small farmers although accessing large, expensive commercial units presents problems.

The Lignozym ® Process in Pulp and Paper.

The Lignozym ® process is discussed in a paper by H. P. Call and I. Mucke (1997) developed at pilot-scale from which much of the discussion is taken below.

Lacasses, one of the major lignases, are produced by white rot fungi (WRF) but has too low a redox potential to oxidize non-phenolic lignin; however WRF that degrade lignin and lack the lignases lignin peroxidase

(LiP) and manganese peroxidase (MnP) has stimulated research on the role of lacasse in lignin degradation (D. O. Krause et al., 2003).

Recent studies with low-molecular-mass 'mediators' such as 2,2'-azino-bis-(3-ethylthiazoline)-6-sulfonate (ABTS) or 3-hydroxy anthranilic acid (3-HAA), lacasse is able to oxidize a wide range of aromatic compounds (D. O. Krause et al., 2003). 3-HAA is a naturally occurring redox mediator and is support for this ligninolytic system as having equivalent potential to ligninolytic systems based on LiP or MnP (D. O. Krause et al., 2003).

Lacasse activity in fungi can be determined with substrates that are specific and sensitive such as syringaldazine; vanillalacetone has also been used (H. P. Call and I. Mucke, 1997).

Recently 3-HAA, ABTS and 1-hydroxybenzotriazole (HBT) have been used in process-scale pulp biobleaching applications; given that enzymes are too large to penetrate the unaltered wood cell wall, these reactive, diffusable low-molecular-mass compounds are responsible for the degradative attack on the lignin polymer; although they remain unidentified, it is likely that lacasses have a preferred low-molecular-mass 'mediator' substrate which represents a major secreted metabolite (H. P. Call and I. Mucke, 1997).

Starting in 1986 Lignozym GmbH (FRG) improved on the effectiveness of the lacasse-mediator-system from the fungi *Trametes (Coriolus) versicolor* with a group of mediators containing N-OH-, N-oxide-, oxime- or hydroxamic-compounds (H. P. Call and I. Mucke, 1997).

To give a background on pulping and bleaching, the conventional process in pulp and paper making, pulping or defribation can be by mechanical grinding, the sulfite process, kraft (sulfate) and further fine-tuned with extended cooking; the cooking process removes the lignin and hemicellulose reducing the kappa number (a measure of the removal of lignin in pulp) although there is still unmodified, modified and repolymerized lignin which has to be removed by subsequent bleaching sequences; there is also reprecipitated xylan which has to be removed by solubilization in alkali or by enzymatically hydrolyzing xylan which may release lignin making it extractable and there is making residual lignin more accessible for chemicals in the bleaching sequences (H. P. Call and I. Mucke, 1997). Bleaching technologies have changed in the last two decades with oxygen delignification introduced in the 1980s and peroxide- and ozone-stages in the 1990s. Biopulping/ biobleaching using the fungi *Phanerochaete chrysosporium* can partially delignify unbleached kraft pulp; process on a small scale has improved; unfortunately, despite improvements, there are still practical obstacles to broad commercial application (H. P. Call and I. Mucke, 1997).

It has been pointed out that separation of lignin from cellulose in pulp processing is still in its infancy (H. P. Call and I. Mucke, 1997).

The lacasse-mediator-system or Lignozym ® process with pilot plant trials in Baienfurt, FRG reported by Call and Mucke (1997) uses a sequence of steps using batch reactors or quantum mixers typically the lacasse stage (L), the extraction stage (E) with the alkali NaOH which extracts lignin and xylan, the Q stage (Q) with diethylenetriamine pentaacetic acid (DTPA) and sulfuric acid, a bleaching stage, and the peroxide stage (P) with hydrogen peroxide and NaOH (alkaline peroxide), also a bleaching stage; pulp with differing lignin content was used (hardwood, softwood or bagasse) as cooked kraft pulps with or without oxygen pre-treatment.

HBT (a R-N-OH mediator compound and one of the most effective) was used and has the net reaction:

$$\text{lacasse}$$
$$2 \text{ substrate-H2 (mediator)} + O2 \text{ - - ->} 2 \text{ substrate (ox)} + 2H2O$$

with lacasse generating a strongly oxidizing intermediate, the co-mediator, in the presence of oxygen which in turn is the real bleaching agent to lignin; lacasse converts the mediator to the co-mediator (presumably RNO.) which penetrates the fibre performing the reaction towards lignin with co-factors 'just in time' (H. P. Call and I. Mucke, 1997). The electron transfers of lacasse involve extraction of four electrons with the reaction:

$$O2 + 4H+ \ + 4e \text{ ->} 2H2O$$

which is similar to the reaction of oxygen metabolism as with cytochrome-c oxidase in the respiratory chain of aerobic organisms (H. P. Call and I. Mucke, 1997). Residual products of the mediator's reaction are benzotriazole (BT), HBT and polymers of BT/or related substances (H. P. Call and I. Mucke, 1997).

For L the conditions are: dry pulp fed at 100 kg per hr at consistency of 10%, pH 4.5, 2 hr residence time at 45 degrees centigrade and 2 bars oxygen pressure, applying 40 IU of enzyme per g of pulp and mediator dosage of 5, 10 or 20 mg per g of pulp; for reasonable delignification 50 to 25% of the original mediator dosage (20 kg per ton) can be used; with the pulps used delignification of >40% can be obtained with 5 kg HBT per ton; mediator is the main cost factor in the process and also impacts oxygen demand and enzyme inactivation by the bleach active agent of the mediator; E had a consistency of 10%, pH of 11.5, residence time of

1 hr at 60 degrees centigrade, with NaOH regulated via pH control; Q had a 5% consistency, pH 5, residence time of 30 mins and temperature of 60 degrees centigrade and dosage of 0.2% DTPA with sulfuric acid; and P had a consistency of 10%, pH of 11.2, residence time of 3.5 hrs and temperature of 75 degrees centigrade with 3% hydrogen peroxide and NaOH via pH control (H. P. Call and I. Mucke, 1997).

The pilot plant trial involved disintegration and screening of the pulp and entered a 4-stage bleach plant with L-E-Q-P reactors; double wide screen presses were designed with high outlet consistencies with recirculation of filtrates; pulp filtrates were recirculated in three very narrow, nearly closed cycles in counter current flow: the acidic filtrates from L and Q and the alkaline filtrates from E and P; start-up problems had to be solved to obtain a continuous run without interruptions and breakdowns; every hour pulp samples were taken from every stage and analyzed for pulp flow, dry substance, pH, kappa, viscousity and brightness to control the run; strength properties were also determined (H. P. Call and I. Mucke, 1997).

Although the process of using cooked kraft pulps processed to a consistency with further treatment with the Lignozym ® process and bleaching sequences was aimed for paper making with the related parameters of viscousity, brightness, strength and other properties related to paper quality measured, the application to treating bagasse as feed would presumably also involve mechanical grinding, pulping (cooking and defibration), enzymatic treatment and bleaching steps with perhaps the use of the practical approach of urea-ammoniation to add nitrogen (N) for digestion.

It remains to be determined what the costs are to the treatment process of bagasse via the Lignozym ® process and at what price when sold to the small farmer.

It is not possible to predict at this point the increase in digestibility of bagasse from the kraft pulp and Lignozym ® process with bleaching sequences. Lower figures given are as follows. Figures with urea-ammoniation pre-treatment are up to only 10% and for genetically modified lignin, the structural basis of which is not well understood, with a decrease in lignin content of 23%, a figure of only 4% (P. T. Doyle et al., 1986, D. J. Cherney et al., 1990). The proposed use of enzymatically delignifying the forage substrate with figures of 50-70% delignification from the Lignozym ® process obtained depending on enzyme and mediator concentrations used and type of pulp (H. P. Call and I. Mucke, 1997) could be an effectively significant means of chemically removing lignin and exposing the contained polysaccharide cellulose for digestion. *Palo-podrido* or highly digestible decayed hardwood used as animal feed in southern Chile that results from the action of WRF (F. Zadrazil et al., 1995) could help attest to this. Use with lower lignin genetically modified feed material would further contribute to its successful digestibility.

The SSF Process.

SSF with fungal cultures is characterized by the complete or almost complete absence of free liquid with water absorbed or in complexed form with the solid matrix and the substrate suitable for fungi requiring low water activity (R. L. Howard et al., 2003).

The effect of treatment of lignified fibrous crop residues is to improve the nutritive value by increasing biomass protein and degrading the phenolic fraction (lignin) at the expense of some structural carbohydrates (F. Zadrazil et al., 1995). Research has been done on a few hundred strains of WRF on wheat straw with increases in digestibility of 15 to 30 units; another study with other substrates resulted in only a 7% increase in digestibility for bagasse by P. gigantea over 21 days (F. Zadrazil et al., 1995). More research is required to find which strains and to determine conditions most effective for bagasse.

The following are selection criteria for fungi for the degradation of lignin: 1) competitive saprophytic ability in terms of relative visible growth on solid substrate under optimal culture conditions during fermentation, 2) improvement in digestibility found to occur in the early stages of partial delignification and to avoid degradation of the polysaccharide; no toxicity of phenolic compounds from lignin breakdown has been observed, 3) minimum polysaccharide required to sustain lignin degradation, 4) resistance of fungi to environmental stress, 5) non-pathogenic and non-toxic to animals and humans, 6) improvement in palatability, 7) genetically stable and 8) easy to prepare inoculum and convenient handling (F. Zadrazil et al., 1995).

Further manipulation of fungi is required to optimize the process of delignification. Basidiomycetes produce too low a level of lignase for sufficient treatment (D. G. Armstrong and H. J. Gilbert, 1991) and mutation or genetic engineering should be used to produce over-expressing strains including the required 'mediators' or secreted metabolites to bring about delignification. Also genetic engineering is required to improve the Kcat or efficiency of the enzymes as they are too slow and to improve stability (such as thermotolerant strains suitable for the tropics) (D. G. Armstrong and H. J. Gilbert, 1991, F. Zadrazil et al., 1995). Cel (cellulose) negative mutants of WRF have been produced which retain a reduced capacity to modify lignin using co-substrates other than cellulose which is not attacked (F. Zadrazil et al., 1995).

The following are conditions optimized to obtain a uniform fungal treated mass: 1) moisture content has several influences: mass transfer of oxygen and carbon dioxide and heat dissipation with consequences on fungal growth, enzyme activity, accessibility of substrate and rate of product formation; water content of 65-75% is optimal for growth on solid substrate; too much water reduces gas exchange and oxygen accessibility and causes leaching of nutrients and clogged interstitial spaces whereas too little water reduces fungal growth because water tension is high and degree of substrate swelling is low and leads to early sporu-lation, 2) temperature is optimal at 20-30 degrees centigrade and acidic pH is optimal for most fungi; higher temperatures lead to less lignin degradation and decreased digestibility of wheat straw; good growth for *Coprinus fimaterius* was found at 32 to 40 degrees centigrade; alkali-tolerant fungi at pHs between 7 and 9 can be used with urea treatment of straw; a high pH provides pasteurization if needed, 3) agitation with aeration provides mixing to improve oxygen supply to microorganisms and removes gaseous metabolites in the void spaces of the substrate, prevents localized heating due to microbial evolution and helps dissipate additives in the fermenter and mixing is part of continuous operation, automation and scale-up of processes; mixing can change microorganism morphology and result in change in metabolism; too much mixing can lead to shear forces damaging growing hyphae tips of fungi and 4) inoculum level at higher levels favour biomass formation of *P. chrysosporium* during SSF and substrate density or packing via interstitial space available for hyphae can limit biomass development (F. Zadrazil et al., 1995).

A fermenter design for SSF described in Zadrazil et al. (1995) at the Institute of Soil Biology, Braunschweig (FRG) had a capacity of 6.9 cubic meters equivalent to 1.50 tonnes of straw or 3.0 tonnes of wood chip constructed of polyurethane foam panels sandwiched by polyester board with a removable front panel with a raised slatted floor covered first with a gliding net then a drag net; the sequence of events starts with the first reactor chamber with substrate that has been pre-treated, watered and thermally conditioned, with the gas phase; the gas phase is measured and controlled during pre-treatment which lasts 48 to 140 hrs with added carbon dioxide which is inversely related to oxygen content; oxygen also has an effect on the development of competitive organisms; the drag net was attached to a winch for unloading the conditioned substrate in the first reactor chamber; equipment which continues the unloading onto the second chamber at bottom is coupled to an elevator with grinding tooth bars for fragmentation and loosens the mass onto a conveyor to bring substrate to a rotating spawning machine with inoculum either as a 30-50 kg grain spawn or 50 liters liquid inoculum containing mycelium 100 g dry matter per 1000 kg of conditioned straw substrate; the conveyor is driven by equipment that drives them via the conveyor in the second chamber; the innoculated substrate then fills into the second reactor chamber via a conveyor for filling with a carrier to move the conveyor, the third chamber, similar as the first, has a raised slatted floor covered with netting; in the third chamber bulk density is decreased which increases streaming of the gas phase.

Control of temperature and removal of heat is the most important problem of fungal growth in deep layers. Heat exchange is by temperature control with a gas cooling/heating system with sensors in the substrate, humidity (steam) and speed of gas movement; a fan is used for air movement through the substrate with a regulated opening for air movement and gas outlet; gas input and gas output through the substrate is regulated with carbon dioxide and oxygen concentrations measured (F. Zadrazil et al., 1995).

Drawbacks in the process are difficulty controlling temperature and moisture in the fermenter and slow conversion rate due to limitations on diffusion; there is a need to improve the SSF fermenter design to minimize conditions in different layers of substrate and to minimize energy inputs, control systems to optimize air circulation and ensure constant air humidity and minimize water evaporation and minimize movement within the substrate (F. Zadrazil et al., 1995).

Space requirements which are low due to small amounts of water added to the substrate results in lower capital costs and operating costs and a simpler reactor design requires less processing energy and excludes the problem of bacterial contamination without the need for pasteurization in most cases; other costs are inputs of labour, transport and cost of substrate and pre-treatments include costs for labour, chemicals and preparation of inoculum; duration of the process also affects costs; 1-2 weeks from a duration of 4-5 weeks, where digestibility is maximized, is recommended to decrease the loss in organic matter which offsets an increase in digestibility maximizing the total digestible dry matter rather than the digestibility (F. Zadrazil et al., 1995).

Toxicological (chemical and biological) examination of fungal use needs evaluation with biological treated feeds with toxins produced by fungi during fermentation such as afflatoxins, or toxins by contaminating organisms and metabolites from lignin degradation; fungal cultures may be pathogenic to handlers or to animals that consume the feed; there is the possibility that the microbe is also allergenic when eaten, touched or breathed by the consumer especially when it produces spores (F. Zadrazil et al., 1995).

The average increase in digestibility of straw in the reactor was 13.8 units comparable to that obtained with sodium hydroxide and ammonia treatment (F. Zadrazil et al., 1995). Genetic manipulation of the biological agents may improve on this. A combination of this with methods such as steam pre-treatment and urea-ammoniation pre-treatment and using genetically designed low-lignin sugarcane in future can be used with the SSF process to improve on digestibility and bring about improvement in digestibility in bagasse.

The Use of Cellulase and Xylanase as EFEs.

Cellulases and xylanases as EFEs have been found generally to improve the rate but not the extent of digestion (D. O. Krause et al., 2003, Y. Wang and T. A. McAllister, 2002). There are variables to consider when utilizing EFEs with feeds.

EFEs marketed are those with cellulases and xylanases primarily although they also contain secondary activities (amylase, proteinase and pectinase) and they would also contain the multiplicity in enzyme activities needed and preparations may vary with strain, growth substrate and culture conditions (Y. Wang and T. A. McAllister, 2002, K. A. Beauchemin et al., 2003). The blended crude preparations of cellulases and hemicellulases to specified levels are limited by the fact that they may have been tested on certain substrates but not for ruminant feeds and may not be limiting given the cell wall-degrading organisms in the rumen (Y. Wang and T. A. McAllister, 2002).

The method of application is an important factor in the effectiveness of utilizing EFEs. EFEs can be sprayed onto concentrate, forage or total mixed rations (TMRs), top-dressed and introduced directly into the rumen; most effective is spraying onto dry feed (forage or concentrate) which may increase the stability of the enzymes with substrate and/or provide for pre-ruminal hydrolysis (Y. Wang and T. A. McAllister, 2002, K. A. Beachemin et al., 2003). It has been speculated that high moisture feeds would have lower binding capacity dissolving the EFEs in ruminal fluid (Y. Wang and T. A. McAllister, 2002). Association with particles would enhance the fact that EFEs attached to feed particles rather than being fluid-associated have the greatest influence on digestion (Y. Wang and T. A. McAllister, 2002). In a study investigating the effects of prehydrolysis showed that where extensive hydrolysis of feed occurred by spraying and then feed held at 39 degrees centrigrade for 24 hrs and freeze-dried prior to batch incubation compared to spraying feed with EFEs and used directly, limited hydrolysis may promote microbial growth with increased availability of reducing sugars without substantial accumulation of

PC-LCCs on the feed surface which would reduce microbial colonization (Y. Wang and T. A. McAllister, 2002). Also applying enzymes to a smaller portion of the diet produces reduced response; it has been suggested that applied to a larger portion increases the chances of the enzymes enduring in the rumen (K. A. Beauchemin et al., 2003).

Feed type or substrate response to EFEs would also play a role with studies across several enzyme formulations involving various forages indicating the importance of optimization by matching the enzyme product to the forage but we are presumably only considering one feed type here, bagasse. Still variations may exist between enzyme formulations with different bagasse feeds.

Finally, the level of animal productivity as with high-producing dairy cows and growing cattle where fibre digestion is compromised due to low ruminal pH and rapid transit times through the rumen or when digestibility is lower as in dairy cows versus sheep affects improvement of feed digestion by exogenous enzymes when the potential digestibility is not attained with the "loss" captured with their use (K. A. Beachemin et al., 2003).

The cost of EFEs is dropping and may be perhaps used in developing country settings in the future.

Shortened Summary.

Bagasse is the fibrous residue of the milled sugarcane plant comprising about 28% of the cane plant and can be used for animal feeding. It was earlier pointed out that treatment of bagasse with chemicals is expensive but treatment with enzymes and fermentation has great promise although it was not further elaborated on (eg. use of yeast or some other microbial source). This paper discusses the theory of PC-LCCs that limit microbial colonization in the feed particle and limit the extent of digestion in the rumen. EFEs do not add novel enzymatic activities to the rumen and thus are limited in their extent of digestion. Two industrial applications as methods of pre-treatment that address this problem are the use of the lignase, the lacasses, with mediator compounds for delignification in pilot pulp paper applied to forage and the SSF process with WRF that degrade lignocelluloses along with other practical approaches to delignification. We will then discuss the use of EFE preparations and the factors that affect response to EFEs. Bagasse is fibrous and highly lignified and is of low digestibility (35%). PC-LCCs at the surface of feed particles act as a physical and biochemical layer that limit microbial colonization and further attack as supported by evidence that free phenolic acids and soluble carbohydrate complexes inhibit microbial activity and prevent microbial attachment. Cell wall degradation will be limited by the rate of microbial colonization and removal of PC-LCCs to expose the contained polysaccharide. Studies have shown that EFEs are limited in their ability to cleave phenolic compounds from feed particles and that extensive hydrolysis reduces colonization compared to minimally hydrolyzed straw despite the presence of reducing sugars which brings us back to the point that EFEs may by hydrolyzing the polysaccharides of the cell wall and leaving the PC-LCCs on the surface to block colonization. With in situ DM disappearance with EFEs the percentage of phenolics in the residue increases. This is consistent with EFEs increasing only the rate but not the extent of digestion. It has been observed that EFEs are limited in cleaving the PC-LCC esterified linkages with research suggesting that linkages are cleaved by acetylesterase and ferulic acid esterase. Bonds between lignin and hemicellulose are with the residues of the hydroxycinnamic acid ferulic acid and the uronic acid 4-O-methyl-beta-D-glucuronic acid in graminaceous plants and ether bonds between lignin and the carbohydrate moiety and lignin composed largely of aryl-glycerol-beta-aryl ether structures constituting 50% of the monomers. Novel esterase and etherase activities need to be identified and produced along with EFEs currently marketed. Other methods which can be used with EFEs together with the industrial applications mentioned are alkaline pre-treatment which cleaves esterified bonds in the PC-LCC matrix improving microbial access, adhesion and colonization and steam pre-treatment which improves hydrolysis efficiency of EFEs; steam pre-treatment separates out the hemicellulose, cellulose and lignin components of fibre. Use of urea-ammoniation may be optimized for maximal efficiency with use of EFEs. Both urea-ammoniation and steam pre-treatment are considered practical approaches for developing country settings. The processing of bagasse at industrial-scale for a developing country has been proposed in the past for the Philippines with a fermentative process for bagasse. Presumably bagasse can be resold to small farmers although accessing large, expensive commercial units presents problems. The Lignozym® process developed at pilot-scale was discussed in a paper by Call and Mucke (1997) using one of the major lignases, the lacasses produced by WRF; WRF without LiP or MnP can degrade lignin which stimulated research on the role of lacasse in lignin degradation. With low-molecular-mass 'mediators' ABTS and 3-HAA, lacasses are able to oxidize a wide range of aromatic compounds. With 3-HAA, a naturally occurring redox mediator is support for a ligninolytic system equivalent to systems based on LiP and MnP. Lacasse activity can be determined with specific and sensitive substrates such as syringaldazine and also with vanillalacetone. Recently ABTS, 3-HAA and HBT have been used in process-scale biobleaching. Given that enzymes are too large to penetrate the unaltered wood cell wall, these reactive, diffusable low-molecular-mass compounds are responsible for the degradative attack on the lignin polymer; lacasses are likely to have a preferred low-molecular-mass 'mediator' as a major secreted metabolite. Starting in 1986 Lignozym GmbH (FRG) improved on the lacasse-mediator-system from fungi *Trametes versicolor* with mediators with groups N-OH-, N-oxide-, oxime- or hydroxamic-compounds. Pulping or defribration can be by mechanical grinding, the sulfite process and kraft (sulfate) process with further extended cooking; the cooking process removes the lignin and hemicellulose reducing the kappa number (related to reduction in lignin in pulp) although there still remains unmodified, modified and repolymerized lignin which can be removed by subsequent bleaching; there is also reprecipitated xylan which has to be removed by solubilization in alkali or by enzymatic hydrolysis of xylan which can release lignin making it extractable or more accessible to bleaching. Bleaching technologies include oxygen delignifi-cation in the 1980s and peroxide- and ozone-stages in the 1990s. Biopulping/ bio-bleaching of unbleached kraft pulp with *Phanerochaete chrysosporium* can partially delignify unbleached kraft pulp with improved results although there are still practical obstacles to broad commercial applications. The separation of lignin from cellulose in pulp processing is still in its infancy. The Lignozym ® process which we will now discuss was developed in pilot-plant trials in Baienfurt, FRG reported by Call and Mucke (1997) using a sequence of batch reactors typically with L using the the Lignozym ® system, E with the alkali NaOH which extracts lignin and xylan, Q with DTPA and sulfuric acid, a bleaching stage, and P with hydrogen peroxide and NaOH (alkaline peroxide), also a bleaching stage; pulp with differing lignin content was used (hardwood, softwood or bagasse) as cooked kraft with or without oxygen pre-treatments. HBT (a R-N-OH mediator and one of the most effective) was used with the net reaction: 2-substrate-H2 (mediator) + O2-lacasse-> 2 substrate (ox) + 2H2O with lacasse generating a strongly oxidizing intermediate, the co-mediator (presumably RNO.) in the presence of oxygen, the real bleaching agent. The electron transfers of lacasse involve four electrons: O2 + 4H+ + 4e -> 2H2O similar to the reaction of oxygen metabolism. Residual products of the mediator's reaction are BT, HBT and polymers of BT/or related substances. For L the conditions are: dry pulp fed at 100 kg per hr at consistency of 10%, pH 4.5, 2 hr residence time at 45 degrees centigrade and 2 bars oxygen pressure applying 40 IU of enzyme per g of pulp and mediator dosage of 5, 10 or

20 mg per g of pulp; for reasonable delignification 50 to 25% of the original mediator dosage (20 kg / ton) could be used; with pulps used >40% delignification can be obtained with 5 kg HBT per ton; the mediator is the main cost factor in the process and impacts oxygen demand and enzyme inactivation by the mediator; E had a consistency of 10%, pH of 11.5, residence time of 1 hr at 60 degrees centigrade, with NaOH regulated via pH control, Q had 5% consistency, pH 5, residency time of 30 mins and temperature of 60 degrees centigrade and dosage of 0.2% DTPA with sulfuric acid and P had a consistency of 10%, pH of 11.2, residence time of 3.5 hrs and temperature of 75 degrees centigrade with 3% hydrogen peroxide and NaOH via pH control. The pilot plant trial involved disintegration and screening of pulp and entered a 4-stage bleach plant with L-E-Q-P reactors; double-wide screen presses were designed with high outlet consistencies with recirculation of filtrates; filtrates were recirculated in three very narrow, nearly closed cycles in counter current flow: the acidic filtrates from L & Q and alkaline filtrates from E & P. Start-up problems has to be solved to obtain a continuous run without interruptions and breakdowns; every hour samples were taken from every stage and analyzed for pulp flow, dry substance, pH, kappa, viscousity and brightness to control the run; strength properties were also determined. Although the process of using cooked kraft pulps with a consistency with treatment with the Lignozym ® process was used for paper making with the properties related to paper quality measured, the application to bagasse as feed would presumably also involve mechanical grinding, pulping (cooking and defibrization), enzymatic treatment and bleaching steps and with the practical approach of urea-ammoniation to add N for digestion. It remains to be determined at what cost the treatment process of bagasse is via the Lignozym ® process for the small farmer. It is not possible to predict the increase in digestibility of bagasse with the kraft pulps and Lignozym ® process with lower figures of up to only 10% obtained with urea-ammoniation and with genetically modified lignin, the structural basis of which is not well understood, with a decrease in lignin content of 23%, a figure of only 4%. The proposed use of straw treatment and enzymatically delignifying the forage substrate with 50-70% delignification from the Lignozym ® process could be an effective means of chemically removing more of the lignin and exposing the contained polysaccharide resulting in more dramatic increases in digestibility. The *palo-podrido* process from the action of WRF on hardwood is highly digestible and could help attest to this. Use of lower lignin feed material would contribute further to increasing digestibility. SSF is characterized by complete or almost complete absence of free liquid with water absorbed with the solid matrix

for fungi requiring low water activity. SSF treats lignified fibrous crop residues to increase biomass protein and degrade lignin at the expense of some structural carbohydrates. Research with WRF on wheat straw increases digestibility of 15 to 30 units. A study with bagasse resulted only in 7% increase in digestibility with *P. gigantea* over 21 days. More research is required to find which strains and to determine conditions most effective for bagasse. Selection of fungi for the degradation of lignin are according to the following criteria: 1) competitive saprophytic ability, 2) improvement in digestibility in the early stages of partial delignification and to avoid degradation of polysaccharide, 3) minimum polysaccharide to sustain lignin degradation, 4) resistance of fungi to environmental stress, 5) non-pathogenic or non-toxic to animals and humans, 6) improvement in palatability, 7) genetically stable and 8) easy to prepare inoculum and convenient handling. Further manipulation of fungi to optimize the process is required. Basidiomycetes produce too low a level of lignase for sufficient treatment and mutation or genetic engineering should be used to produce over-expressing strains including the required 'mediators' or secreted metabolites to bring about delignification. Genetic engineering is also required to improve the Kcat as enzymes are too slow and to improve stability such as with tropical thermotolerant strains. Cel negative mutants of WRF have been produced to retain a reduced capacity to modify lignin using co-substrates other than cellulose. The following are conditions optimized to obtain a uniform fungal treated mass: 1) moisture content which influences mass transfer of oxygen and carbon dioxide and heat dissipation with consequences on fungal growth, enzyme activity, accessibility of substrate and rate of product formation; 65-75% water content is optimal for growth on solid substrate; too much water reduces gas exchange and oxygen accessibility and causes leaching of nutrients and clogged interstitial spaces whereas too little water reduces fungal growth because water tension is high and degree of substrate swelling is low and leads to early sporulation, 2) temperature is optimal at 20-30 degrees centigrade and acidic pH is optimal for most fungi; high temperature leads to less lignin degradation and decreased digestibility of wheat straw; good growth for *Coprinus fimaterius* was found at 32 to 40 degress centigrade; alkali-tolerant fungi between pH of 7 to 9 can be used with urea treatment of straw; a high pH provides for pasteurization and 3) agitation with aeration provides mixing to improve oxygen supply to microorganisms and removes gaseous metabolites in the void spaces of the substrate, prevents localized heating and helps dissipate additives and mixing is part of continuous operation, automation and scale-up; mixing can change microbial morphology and change metabolism in fungi; too much mixing can lead to shear forces damaging hyphae tips of fungi and 4) inoculum levels with higher levels favouring biomass formation of *P. chrysosporium* and substrate density or packing via interstitial space available for hyphae can limit biomass development. A fermenter design for SSF described by Zadrazil et al. (1995) had a capacity of 6.9 cubic meters or 1.50 tonnes of straw or 3.0 tonnes of wood chip constructed of polyurethane foam panels sandwiched by polyester board with a removal front panel with a raised slatted floor covered first with a gliding net then a drag net; the sequence of events starts with the SSF reactor chamber with substrate that has been watered and thermally conditioned with the drag net attached to a winch for unloading; the gas phase is also measured and controlled during the first stage of pre-treatment which lasts 48 to 140 hrs with carbon dioxide which is inversely related to oxygen; oxygen has an effect on the development of competitive organisms; equipment which continues the unloading to the second chamber at bottom is coupled to an elevator with grinding tooth bars for fragmentation and loosens mass onto a conveyor to bring the substrate to a rotating spawning machine with inoculum either as a 30-50 kg grain spawn or 50 liter liquid innoculum containing mycelium 100 g dry matter per 1000 kg of conditioned straw substrate; the conveyor is driven via equipment in the second chamber; the innoculated substrate then fills a third chamber via a conveyor with carrier to move the conveyor; the third chamber, similar as the first, has a raised slatted floor covered with netting; in the third chamber bulk density is decreased which increases streaming of the gas phase. Control of temperature and removal of heat is the most important problem of fungal growth in deep layers. Heat exchange is by temperature control with a gas cooling/heating system with sensors in the substrate, humidity (steam) and speed of gas movement; a fan is used for air movement through the substrate; there is also a regulated opening for gas movement and a gas outlet; gas input and gas output through the substrate is regulated with carbon dioxide and oxygen concentrations measured. Drawbacks in the process are difficulty in controlling temperature and moisture in the fermenter and slow conversion rate due to limitations on diffusion. There is a need to improve the SSF fermentor design to minimize energy inputs, control systems to optimize air circulation, constant air humidity and minimize water evaporation and minimize movement within the substrate. Space requirements which are low due to small amounts of water added results in lower capital and operating costs and a simpler reactor design requires less processing energy and without the need for pasteu-rization in most cases; other costs are inputs of labour, transport and costs of substrate and pre-treatment costs of labour, chemicals and preparation of inoculum;

duration of the process also affects costs: 1-2 weeks from a duration of
4-5 weeks where digestibility is maximized, to maximize total digestible
dry matter rather than digestibility. Toxicological examination of fungal
use needs evaluation with biological treated feeds with toxins such as
afflatoxins produced by fungi or metabolites of lignin degradation;
fungal cultures may be pathogenic to handlers or animals that consume
the feed; there is the possibility that the microbes are also allergenic
when eaten, touched or breathed by the consumer especially when it
produces spores. The average increase in digestibility of straw in the
reactor was 13.8 units comparable to that obtained with sodium hydroxide
and ammonia treatment. Genetic manipulation may improve on this. A
combination   of   this   with   methods   of   straw   urea-ammoniation
pre-treatment and use of genetically designed low-lignin sugarcane in
future can be used with the SSF process to bring about a significant
improvement in digestibility. Cellulases and xylanases as EFEs have been
found to improve the rate but not extent of digestion. There are variables
to consider when utilizing EFEs with feeds. EFEs marketed contain
cellulases and xylanases presumably with other secondary activities with
a multiplicity in enzyme activities and preparations may vary with strain,
growth substrate and culture conditions. The blended crude preparations
of cellulases and hemicellulases are limited by the fact that they may
have been tested on certain substrates but not for ruminant feeds and
may not be limiting given the cell wall degrading organisms in the
rumen. Methods of application of EFEs can be by spraying onto
concentrate, forage or TMR, top-dressed and introduced directly into
the rumen; most effective is by spraying onto dry feed which may increase
the stability of the enzymes with substrate and/or provide for pre-ruminal
hydrolysis with high moisture feeds having a lower binding capacity;
being attached to feed particles rather than fluid have the greatest
influence on digestion; in a study investigating the effects of pre-hydrolysis
with extensive hydrolysis at 39 degrees centigrade for 24 hrs and
freeze-dried and compared with spraying and direct use, the latter may
promote microbial growth with increased availability of reducing sugars
without substantial accumulation of PC-LCCs on the feed surface
reducing colonization; also applying EFEs to a larger portion of the feed
increases the chances of enzymes enduring in the rumen. Feed type
would also play a role in response to EFEs and variations may exist
between different bagasse feeds. Finally, the level of animal productivity
as with high-producing dairy cows and growing cattle due to low ruminal
pH and rapid transit times through the rumen or when digestibility is
lower as in dairy cows versus sheep affects improvement of feed digestion
by exogenous enzymes when the potential digestibility is not attained

and this "loss" is captured. Cost of EFEs is dropping and may be used in developing countries in the future.

References.

1. D. G. Armstrong and H. J. Gilbert. 1991. The application of biotechnologies for future livestock production. In: Physiological Aspects of Digestion and metabolism in Ruminants. T. Tsuda, Y. Sasaki and R. Kawashima (Eds.). Pp. 737-761. Academic Press. San Diego, USA.

2. E. T. Baconawa. 1986. Case study-Prospects for reconversion of sugarcane into animal feed in the Philippines. In: Sugarcane as Feed. R. Sansoucy, G. Aarts and T. R. Preston (Eds.). FAO Animal Production and Health Paper 72. FAO-UN Rome, Italy.

3. K. A. Beauchemin, D. Colombatto, D. P. Morgavi and W. Z. Yang. 2002. Use of Exogenous Fibrolytic Enzymes to Improve Feed Utilization by Ruminants. Journal of Animal Science 81 (E suppl.): E37-E47.

4. H. P. Call and I. Mucke. 1997. Minireview. History, overview and applications of mediated lignolytic systems, especially laccase-mediator-systems (Lignozym ® -process). J. of Biotechnology 53: 163-202.

5. D. J. Cherney, J. A. Patterson and K. D. Johnson. 1990. Digestibility and feeding value of pearl millet as influenced by the brown-midrib, low-lignin trait. Journal of Animal Science 68: 4345-4351.

6. P. T. Doyle, C. Devendra and G. R. Pearce. 1986. Improving the feeding value through pretreatments. In: Rice Straw as a Feed for Ruminants. Pp. 54-89. IDP Canberra Australia.

7. K. E. Hammel. 1996. Extracellular free radical biochemistry of ligninolytic fungi. New J. of Chemistry 20: 195-198.

8. R. L. Howard, E. Abotsi, E. L. Jansen van Rensburg, S. Howard. 2003. Ligno- cellulose biotechnology: issues of bioconversion and enzyme production. African J. of Biotechnology 2: 602-619.

9. D. O. Krause, S. E. Denman, R. I. Mackie, M. Morrison, A. L. Rae, G. T. Attwood and C. S. McSweeney. 2003. Opportunities to improve fibre degradation in the rumen: microbiology, ecology and genomics. FEMS Microbiology Reviews 27: 663-693.

10. E. R. Orskov. 2002. Crop Fractionation. In: Trails and Trials in Livestock Research. Pp. 67-69. International Feed Resources Unit, Aberdeen, U. K.

11. Y. Wang and T. A. McAllister. 2002. Rumen microbes, enzymes and feed digestion. Internet document.

12. F. Zadrazil, A. K. Puniya and K. Singh. 1995. Biological upgrading of feed and feed components. In: Biotechnology in Animal Feeds and Animal Feeding. R. J. Wallace and A. Chesson (Eds.). Pp. 55-70. VCH Weinheim FRG.

Overview

A monograph collection on various topics with implications on new or possible advances with ligno-cellulose research in regards to animal feeding and also lending itself to bioenergy feedstock. It is an informative discussion for the research scientist, and in particular, the specialist in ruminant nutrition covering such topics as follows. Enzyme technology, applied to crop post-harvest technology, with novel microbial anaerobic lignases, aerobic lignases and other extracellular fibrolytic enzymes (EFEs), boosting water-soluble carbohydrate (WSC) content in new tropical forage-type feeds, action of proteases in plant feed material and digestion, lowering lignin content and use of lacasse for bio-bleaching lignocellulose, as examples. Feed resources discussed, in particular in Asia, including sugarcane and use of bagasse and tops, grasses and legumes, with resources for food and feed farming systems and legume browse tree and shrubs for feed. The issues of various pre-treatments and crop improvements with biotechnology and digestion are discussed.